W9-ASF-143

ALLE ZEIT WACH
1842

H.-O. Peitgen P. H. Richter

# The Beauty of Fractals

## Images of Complex Dynamical Systems

With 184 Figures, Many in Color

Springer-Verlag
Berlin Heidelberg New York Tokyo

Prof. Dr. Heinz-Otto Peitgen
Fachbereich Mathematik, Universität Bremen, D-2800 Bremen 33
and Department of Mathematics, University of California, Santa Cruz
Santa Cruz, CA 95064, USA

Prof. Dr. Peter H. Richter
Fachbereich Physik, Universität Bremen, D-2800 Bremen 33

The cover picture is a 3-dimensional rendering of the electrostatic potential of the satellite Mandelbrot set from Map 27. A purely mathematical construct, this landscape bridges the gap between real and artificial fractals. The "moon" is a Riemann sphere which carries the structure of a corresponding Julia set. As a symbol of Euclidean geometry, the sphere contrasts strongly with the fractal nature of the mountain ridge. (Computer graphics by Dietmar Saupe)

ISBN 3-540-15851-0 Springer-Verlag Berlin Heidelberg New York Tokyo
ISBN 0-387-15851-0 Springer-Verlag New York Heidelberg Berlin Tokyo

Library of Congress Cataloging-in-Publication Data. Peitgen, Heinz-Otto, 1945–. The beauty of fractals. Bibliography: p. Includes index. 1. Fractals. I. Richter, P.H. (Peter H.), 1945–. II. Title. QA447.P45 1986 516 86-3917



Typesetting, Printing, and Binding: Appl, Wemding. 2141/3140-5

# PREFACE

*In 1953 I realized that the straight line leads to the downfall of mankind. But the straight line has become an absolute tyranny. The straight line is something cowardly drawn with a rule, without thought or feeling; it is the line which does not exist in nature. And that line is the rotten foundation of our doomed civilization. Even if there are places where it is recognized that this line is rapidly leading to perdition, its course continues to be plotted . . . Any design undertaken with the straight line will be stillborn. Today we are witnessing the triumph of rationalist knowhow and yet, at the same time, we find ourselves confronted with emptiness. An esthetic void, desert of uniformity, criminal sterility, loss of creative power. Even creativity is prefabricated. We have become impotent. We are no longer able to create. That is our real illiteracy.*

*Friedensreich Hundertwasser*

Fractals are all around us, in the shape of a mountain range or in the windings of a coast line. Like cloud formations and flickering fires some fractals undergo never-ending changes while others, like trees or our own vascular systems, retain the structure they acquired in their development. To non-scientists it may seem odd that such familiar things have recently become the focus of intense research. But familiarity is not enough to ensure that scientists have the tools for an adequate understanding. A child is familiar with his blue cradle and the blue sky long before he is conscious of blue as a common quality of different items. In his cognitive development there is a stage when he becomes receptive to the notion of color; he hears that the sky is blue and suddenly "discovers" that other things are blue too.
The development of our scientific perception of the world follows a similar pattern. Yes, many fractals are familiar to us, but until very recently there was no place for them in our scientific view of nature. That view was shaped by Galileo Galilei, whose mastery of counterintuitive abstraction set the example for modern scientific reasoning. His credo, stated in his own words of 1623:

*Philosophy is written in this grand book – I mean the Universe – which stands continuously open to our gaze, but it cannot be understood unless one first learns to comprehend the language in which it is written. It is written in the language of mathematics, and its characters are triangles, circles and other geometric figures, without which it is humanly impossible to understand a single word of it; without these, one is wandering about in a dark labyrinth.*

It took some 350 years to overcome Galileo's restraint: until Benoit Mandelbrot developed the notion of a fractal. Looking back in 1984, he pondered:

*Why is geometry often described as cold and dry? One reason lies in its inability to describe the shape of a cloud, a mountain, a coastline, or a tree. Clouds are not spheres, mountains are not cones, coastlines are not circles, and bark is not smooth, nor does lightning travel in a straight line . . . . . . Nature exhibits not simply a higher degree but an altogether different level of complexity. The number of distinct scales of length of patterns is for all purposes infinite.*
*The existence of these patterns challenges us to study those forms that Euclid leaves aside as being formless, to investigate the morphology of the amorphous. Mathematicians have disdained this challenge, however, and have increasingly chosen to flee from nature by devising theories unrelated to anything we can see or feel.*

The mathematical concept of a fractal characterizes objects with structures on various scales, large as well as small, and thus reflects a hierarchical principle of organization. There is one important idealization involved: fractal objects are *self-similar,* i.e. they do not change their appearance significantly when viewed under a microscope of arbitrary magnifying power. Though this may be oversimplifying, it does add a dimension of depth to our mathematical representation of nature. Mandelbrot's studies were highlighted by his discovery, in 1980, of the set which now bears his name. He found a principle that organizes a whole universe of self-similar structures in an unexpected way. The bizarre shape that appears on the cover of this book may turn out to be a key element of a new "natural" mathematics, just as the straight line is a constitutive element of Euclidean geometry.

Perhaps the most convincing argument in favor of the study of fractals is their sheer beauty. Through this book we intend to share our own fascination with a general audience. We are encouraged to do so by the overwhelmingly positive response that our pictures have found so far in a series of exhibitions, first within Germany, and recently abroad under the auspices of the Goethe Institute.

Our pictures do not attempt to mimic nature. There are others whose computer graphics have been very successful in that respect, to the extent that in some popular motion pictures artificial landscapes have replaced the real ones. The special fractals shown here have been chosen for their significance in our own scientific work. As the accompanying essays show, they are part of an ongoing worldwide activity to unravel some of the secrets of complex dynamical systems.

When we, mathematicians and physicists at the University of Bremen, first held a workshop on Chaotic Dynamics in the fall of 1981, we certainly could not foresee where our collaboration would lead eventually. But one thing was clear: Insight into the complexity of non-linear systems was to be gained mainly by computer experimentation.

For this we needed appropriate equipment. The young University of Bremen, still expanding, gave us all the support possible. Moreover, some of us were spending time at the University of Utah in Salt Lake City, and there we became acquainted with the excellent equipment of the Computer Science Department. With the help of a grant from the Volkswagen Foundation, we were able to plan and establish our own computer graphics laboratory "Dynamical Systems", and it was here in Bremen that all the pictures of our exhibition were produced.

While we were discussing the slides and videos brought home from Salt Lake City, the idea for a public exhibition came up. These pictures showed the dynamics of area-conserving maps with which our collaboration had begun, and fascinating Julia sets generated by Newton's root finding algorithm.

The idea reached its final shape when Senator *Rolf Speckmann* and *Hans-Christian Bömers* from the Sparkasse in Bremen, both well known for their active support of the arts and sciences, invited us to present our pictures to a larger public in the stately main hall of their bank.

Leaving our protected ivory tower for this first exhibition proved a unique learning experience for us. We found ourselves confronted with challenges

very different from those to which we were accustomed in our professional lives. This was particularly true for the catalogue in which the scientific background of the pictures was to be explained to the general public. Without any experience in such matters we produced the brochure “Harmonie in Chaos und Kosmos”, a work which found more readers-even amongst our colleagues-than any of our purely scientific publications ever had.
The success of the first exhibition in January 1984 put us under pressure. To reach a larger audience, the exhibition had to be developed and expanded. It was important for us at this point that our former teachers, *Friedrich Hirzebruch,* Bonn, and *Manfred Eigen,* Göttingen, who had supported our rather odd activities, invited us to present our material at their Max-Planck-Institutes. These invitations prompted us to set up an almost entirely new exhibition with another catalogue “Morphologie komplexer Grenzen”, all no more than four months after our first Bremen adventure!
The third generation of our pictures has been newly organized as the exhibition “Schönheit im Chaos/Frontiers of Chaos”, in collaboration with the Goethe Institute. We are happy to have been given this opportunity to go abroad, and we are grateful for the active support a good number of friends have given us. *Arnold Mandell,* a psychiatrist in San Diego, established important connections for us when he transmitted his own fascination with our pictures to Deputy Director *Thomas K. Seligmann* at the Fine Arts Museum in San Francisco, and to the late Director of the Exploratorium, *Frank Oppenheimer.* It was in the fall of 1984 that the Goethe Institute added us to their cultural program, and for this we are especially indebted to *Manfred Triesch* of the San Francisco branch, and to *Fritjof Korn* and *Manfred Broenner* at the headquarters in Munich.
In its original form, this exhibition included not only CIBACHROME prints, light boxes and a video film but also a catalogue of 108 pages in German and English which reflected our wanderings between science and art to a greater degree than had the two previous catalogues. The first printing was sold out after a few months. We are grateful to Springer-Verlag for showing such faith in us and producing this book to document and accompany the exhibition. In this book we have added a number of color plates not in the catalogue, and expanded the discussion of the scientific context.
Our essays “Frontiers of Chaos” and “Magnetism and Complex Boundaries” explain the background to the non-specialist; we have therefore had to give up mathematical rigor. Sometimes this proves a hindrance, for it is hard to be clear on what part is actually fact and what part is insight suggested by the experimental results. This is particularly annoying where a rigorous discussion could lead to the heart of the problem. In order to address this shortcoming we have included ten special sections. These sections require a certain familiarity with mathematical and physical thinking. The layman may simply ignore them while students of complex dynamics might find them useful as reference texts on the respective topics.
We are happy that four of our most distinguished colleagues agreed to contribute to this volume with original articles of their own. In different ways they have each been influential in the development of the field, and their reflections add greatly to the authenticity of this volume.

*Behind:*
Hartmut Jürgens, Peter H. Richter, and Heinz-Otto Peitgen

*In front:*
Michael Prüfer and Dietmar Saupe

*Benoit B. Mandelbrot* reports on his discovery of the Mandelbrot set that plays such a central role in the exhibition. He also gives a rather personal account of the history of fractal geometry and of his very special role as a nomad-by-choice, wandering between mathematics and the other sciences.
*Adrien Douady,* one of the leading experts on the mathematics of the Mandelbrot set, recounts what is known and what is still mysterious about this paradigm of beautiful complexity. The more technical passages of his article have a natural overlap with some of our special sections; we hope the interested reader will find that helpful rather than annoying.
*Gert Eilenberger,* the physicist, describes the symbolic meaning that our pictures may have within the changing comprehension of nature. His friendly criticism has accompanied our activities over the years, and has been a welcome source of support.
*Herbert W. Franke,* one of the pioneers of computer graphics, reports on his own experiences and draws a number of inferences from them.
The pictures selected for this volume are essentially those in the exhibition "Frontiers of Chaos". They were chosen from several hundred experiments carried out in 1984 in our laboratory in Bremen. Often we show the very same experiment in different coloration. Having different colors at our disposal has proved a very important tool for looking into complex structures. The selection and composition of colors is arbitrary in a way similar to the orchestration of a musical piece: although the specific musical instrument chosen is not in itself essential to the basic idea, it has a great effect on the experience of the listener. For that reason the choice of color must be done with

great care. In this regard, our graphics can be considered the results of individual and creative acts, as well as authentic and stringent mathematical experiments.
For all those who are less interested in viewing *our* pictures than they are in producing their own, we add some relevant information at the very end of the book.
The translation of German into English was done by *Howard* and *Irene Schultens*. We thank *Nancy Hollomon* and *Eugen Allgower* for critically reading the manuscript.
It is impossible to mention all those who contributed to the exhibition and to this book by being helpful or, at least, tolerant. Above all, we are indebted to our coworkers in the "Graphiklabor Dynamische Systeme", *Hartmut Jürgens, Michael Prüfer, Dietmar Saupe,* and to one of our students, *Heinz-Werner Ramke.* At the Bremen University Computing Center, we were helped especially by *Georg Heygster* in the printing of black-and-white laser graphics. We also enjoyed the valuable sponsorship of the friendly Bremen environment, both its people and business. The University and the Senate of the State of Bremen gave us their full support, and we benefitted from the encouragement of a number of scientific colleagues.
We would particularly like to thank *Sir Michael Atiyah* of Oxford University for his initiative and his many ideas during the planning stage that made the series possible in the UK and catalyzed the international echo for it.
Last but not least we express our gratitude for the inexhaustible patience of our families.

We dedicate this book to Karin and Christiane.

Bremen, Mai 1986 H.-O. Peitgen and P. H. Richter

*Peter H. Richter*

*1945 in Fallingbostel. Studied physics in Göttingen and Marburg. Dr. rer. nat. 1971 as student of S. Großmann. Research on statistical physics of non-equilibrium systems in Göttingen, Cambridge (Mass.), and Stanford. 1980 Professor of Physics at the University of Bremen. 1985/86 Visiting Professor at Boston University.

*Heinz-Otto Peitgen*

*1945 in Bruch. Studied mathematics, physics and economics in Bonn. Dr. rer. nat. 1973 at Bonn. Research on nonlinear analysis and dynamical systems in Bonn and Salt Lake City. 1977 Professor of Mathematics at the University of Bremen. 1985 Professor of Mathematics at the University of California at Santa Cruz. Visiting Professor in Belgium, Italy, Mexico and USA.

*Benoit B. Mandelbrot*

*1924 in Warsaw. Moved to Paris in 1936, 1945–47 Ecole Polytechnique. 1948 M.S. Aerosciences at Caltech, Pasadena. 1952 Ph.D. in Mathematics, University of Paris. Visiting scientist in Princeton, Geneva, Paris, before he moved to the US in 1958. IBM research staff member and IBM Fellow (1974) in Yorktown Heights, N.Y. 1984 Professor for the Practice of Mathematics, Harvard University.

*Adrien Douady*

*1935 in Grenoble. Ecole Normale Supérieure, Paris. Agrégation 1957, Doctorat 1965. Professor of Mathematics at the University of Nice (1965–1970), thereafter at Paris-Sud (Orsay) and at the Ecole Normale Supérieure. Research on Complex Analytic Geometry. Became interested in holomorphic dynamics under the influence of his former student J. H. Hubbard, and also of D. Sullivan, in 1980.

*Gert Eilenberger*

*1936 in Hamburg. Dr. rer. nat. 1961 at Göttingen as student of F. Hund. Research in solid state theory, particularly superconductivity, and nonlinear phenomena. 1970 Professor at the University of Cologne and Director at the Institut für Festkörperforschung, KFA Jülich.
*Convinced that the rationality of science, expanded properly, is the sole and all-embracing source of cognition for mankind, the only religion of an enlightened future.*

*Herbert W. Franke*

*1927 in Vienna. Ph.D. 1950 in Vienna. Industrial research, free-lance journalist. In 1955 he became involved in rational aesthetics, and from 1961 on established himself as Germany's best known science fiction author. Since 1970 he has been producing computer graphics on digital systems. Franke is one of the foremost advocats of technically assisted art.

# CONTENTS

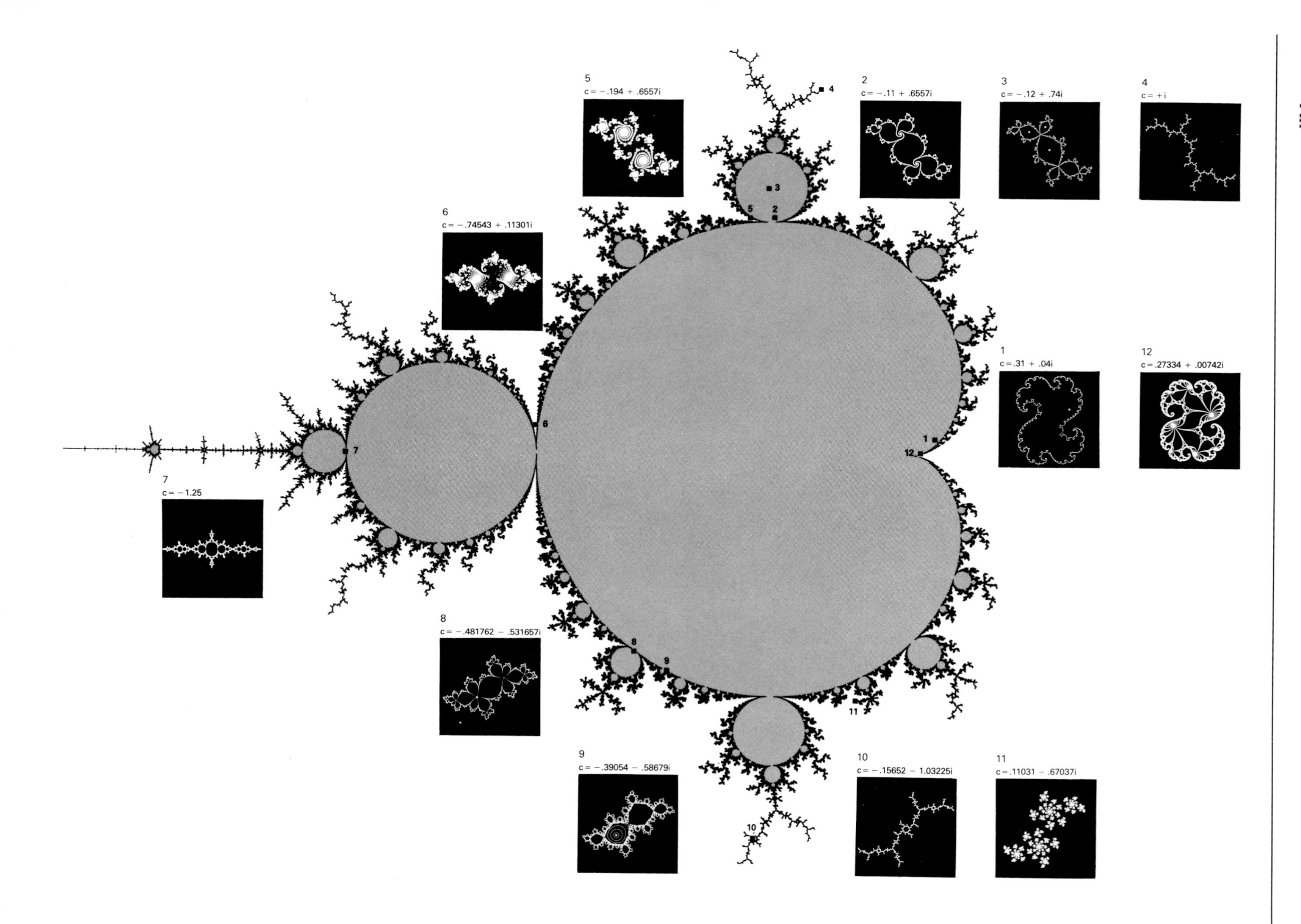

5
c = −.194 + .6557i
2
c = −.11 + .6557i
3
c = −.12 + .74i
4
c = +i
6
c = −.74543 + .11301i
1
c = .31 + .04i
12
c = .27334 + .00742i
7
c = −1.25
8
c = −.481762 − .531657i
9
c = −.39054 − .58679i
10
c = −.15652 − 1.03225i
11
c = .11031 − .67037i

# FRONTIERS OF CHAOS

*Where the world ceases to be the stage*
*for personal hopes and desires, where we, as free beings,*
*behold it in wonder, to question and to contemplate,*
*there we enter the realm of art and of science.*
*If we trace out what we behold and experience*
*through the language of logic, we are doing science;*
*if we show it in forms whose interrelationships are not*
*accessible to our conscious thought but are*
*intuitively recognized as meaningful, we are doing art.*
*Common to both is the devotion to something*
*beyond the personal, removed from the arbitrary.*

*A. Einstein*

*Science and art:* Two complementary ways of experiencing the natural world – the one analytic, the other intuitive. We have become accustomed to seeing them as opposite poles, yet don't they depend on one another? The thinker, trying to penetrate natural phenomena with his understanding, seeking to reduce all complexity to a few fundamental laws – isn't he also the dreamer plunging himself into the richness of forms and seeing himself as part of the eternal play of natural events?

This experience of oneness which the individual may feel finds no counterpart in the intellectual history of the last two hundred years. As if they felt too confined in one soul, the mind of art and the mind of science have parted: *One* Faust has become two one-dimensional beings. The divergence seems irreversible, and what both sides promoted together during the enlightenment is now out of balance. The courage to use one's own reason has turned to presumptuousness. The cool rationality of science and technology has pervaded and transformed the world to such an extent that it could destroy human life. The inspiration of the arts can only respond helplessly, with bitterness.

Of course, this tension affects the natural sciences. Many mature thinkers are becoming aware of the inadequacies of established wisdom. In spite of great, unifying successes in elementary particle physics or sequence analysis in molecular genetics, the creed of the "fundamentalists" has lost its exclusive attractiveness. It is no longer sufficient to discover basic laws and understand how the world works "in principle". It becomes more and more important to figure out the patterns through which these principles show themselves in reality. More than just fundamental laws are operating in what actually *is.* Every nonlinear process leads to branch points, to forks in the path at which the system may take one branch or another. Decisions are made whose consequences cannot be predicted, because each decision has the character of an amplification.

◁ *Julia sets surrounding the Mandelbrot set which controls their structure. The Julia sets corresponding to points inside the Mandelbrot set are connected. They disintegrate into sets of isolated points as the parameter crosses the boundary of the Mandelbrot set. (The Mandelbrot set was kindly provided by Prof. J. Milnor, Princeton)*

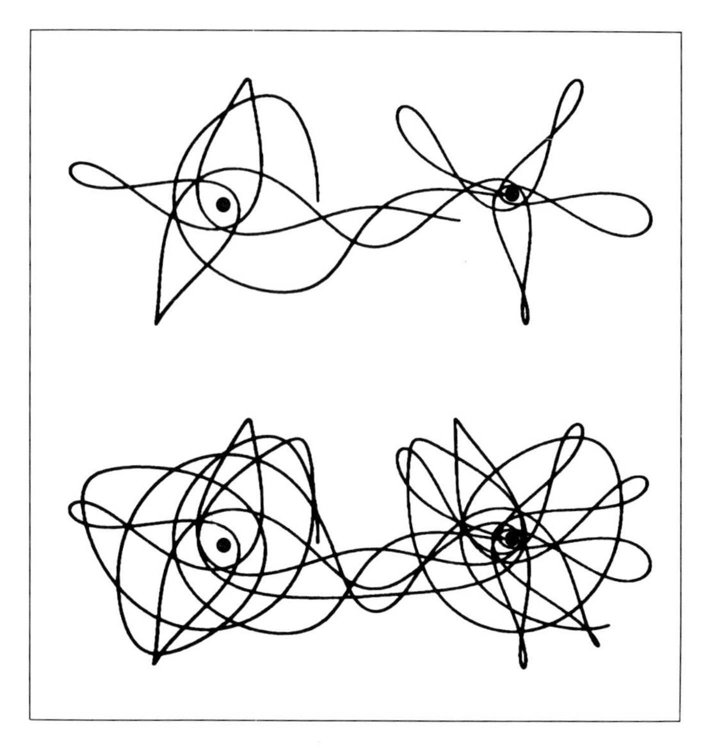

*Fig. 1.* Left: *Typical orbit in a three body problem of celestial mechanics. The upper part shows the beginning, the lower part the sequel of the chaotic motion of a small planet around two suns of equal mass.* Right: *The Lorenz attractor: chaotic motion in a dissipative system. In contrast to planetary motion, this motion is driven towards an attractor by the action of friction. But "strange attractors" allow for complex motion that is jumping back and forth between two centers*

The smallest differences are blown up and have far-reaching effects. Causality holds at every single instant, but it does not carry over a sequence of branchings. Sooner or later the initial knowledge of the system becomes irrelevant. In the unfolding of a process, information is generated and retained. The natural laws allow for many different courses of events, but our world has only one history.

Even in the oldest of natural sciences, astronomy, old appraisals must be revised. When Kepler and Newton, and Einstein more precisely, explained once and for all how single planets move around the sun in elliptical orbits, they gave the impression that merely greater technical effort in calculation would give a complete explanation of the motion of a system of three or more bodies. It is true that our space ships can trust Newton's laws of motion and modern computers to guide them to their goals, but the fact remains that the route becomes unpredictable over *very long periods of time.* The old problem of the stability of the solar system is still unsolved. Around 1800 stability was thought to be proven. Around 1900 – after Poincaré – there were indications to the contrary. Today we have to admit that a prognosis about the long-time behavior of the solar system (even when we restrict the problem to gravitational interactions) is not possible: the equations are "not integrable", as the

expert says. A very small imprecision in the initial conditions can grow to an enormous effect in the later motion. Both the expert and the layman are confused by the complexity contained in what were thought to be simple equations (Fig. 1).
Analogous problems arise in almost all other disciplines. One reason that we have not yet achieved controlled nuclear fusion is that we do not have an adequate understanding of the chaotic motion of a charged particle in the magnetic mirror system. And as the study of developing insect eggs has shown, morphogenesis cannot be understood from just the knowledge of the relevant genome and the molecular machinery. Phenomenology has its own laws. At every new stage of organization new rules take effect.
This is not to imply that the laws of nature known up to now are invalid; it is just difficult to discover all that is hidden in them. This difficulty is common to celestial mechanics, elementary particle physics, developmental biology, and economics. We know it well from everyday life, but it calls for a completely new orientation in science. Basic science must change its view from downward to upward, from the basis to the phenomenon. It needs new paradigms, models which show the heart of the difficulty and point to new avenues of thought. "World models" which drag in hundreds of equations for the discussion of specific questions are not sufficient. They obscure the issue they should illuminate. Knowledge grows from the struggle to find the essential elements and present them "in a nutshell".

## *Thinking in Pictures*

How can this be done if Faust was right in his diagnosis?

*Mysterious in the light of day,*
*Nature retains her veil, despite our clamors:*
*That which she doth not willingly display*
*Cannot be wrenched from her with levers, screws, and hammers.*

... but maybe with the devil's help?

The computer is looked upon as a diabolical instrument by many, scientists no less than artists and worried parents. Some, after a brief glance at the machine, find themselves completely addicted. There must be a reason.
In actual fact, this new medium is allowing us to see connections and meanings which were hidden until now. Especially interactive computer graphics, currently under intensive development, is enriching our perception to a degree rarely achieved by any tool in science. Certainly, it can present us with imaginary worlds, put us into artificial landscapes, and cause us to forget the real world. But used with some reflection, it can also help us lift the veil on nature's secrets.
Where scientists of earlier generations had to simplify their equations drastically or give up completely, we are able to see their full content on the display monitor. In graphical representation, natural processes can be comprehended in their full complexity by intuition. New ideas and associations are

stimulated, and the creative potential of all those who think in pictures is awakened.
Mathematicians, and physicists in particular, have always thought in pictures; even more, they use aesthetic categories as criteria at least for completeness, if not for truth. Hermann Weyl, one of the most important German mathematicians, whose 100th birthday we celebrated in 1985, admitted freely:

*"My work has always tried to unite the true with the beautiful and when I had to choose one or the other I usually chose the beautiful."*

A deep feeling for the unity of science and art speaks out of statements like this. The responses to our pictures quite often express the hope that this unity will become more easily visible, not just as a remote kind of beauty accessible to a few initiates, as is the case, for instance, with Einstein's theory of gravitation. Trivialisation? Perhaps, but this would not be the first time that a craft advanced the cause of a higher spirit.
A transformation has already begun. The separation of traditional disciplines is becoming indistinct. Centers for the study of "Complex Dynamics", "Nonlinear Phenomena" and similar topics have sprung up, prudently not indicating if they are dealing with phenomena from physics, chemistry, biology, or from completely different fields. In the seminars of these centers, the metamorphosis of plants and animals, as well as problems of plasma physics, perceptual psychology, or social behavior are studied. The conviction is growing that formation processes and self-organization develop according to a few typical scenarios which are independent of the specific system involved. In West Germany, for example, Hermann Haken at the University of Stuttgart has been working since the late sixties to establish the discipline of "Synergetics". As a founder of laser theory he discovered that the internal structure formation in the laser takes place according to principles very similar to the competition of molecular species which Manfred Eigen (Max-Planck-Institut in Göttingen) described in his analysis of the early evolution of life. Synergetics is systematically trying to find the rules by which order arises in complex systems.
Our pictures are part of this tradition. They deal with chaos and order, and with their competition or coexistence. They show the transition from one to another and how magnificently complex the transitional region generally is. And though the pictures revel in the beauty of these regions, they are also an attempt to understand the central question of how the structure of the boundaries depends on parameters. This leads us to new boundaries at another level and reveals regularities that no one had suspected a few years ago.
The processes chosen here come from various physical or mathematical problems. They all have in common the competition of several centers for the domination of a plane. A simple boundary between territories is seldom the result of this contest. Rather there is unending filigreed entanglement and unceasing bargaining for even the smallest areas.
It is the border region where the transition from one form of existence to another takes place: from order to disorder, magnetic to non-magnetic state, or however the entities which meet at the boundary are to be interpreted. The border regions are more or less meshed depending on the conditions charac-

terizing the process in question. Occasionally, a third competitor profits from the dispute of two others and establishes its own area of influence. It can happen that one center dominates the entire plane; but there are still "boundaries" of its power, in the form of isolated points which are not subjected to its attraction. These are dissidents, so to speak, who don't want to belong.
The pictures represent processes which are, of course, simplified idealizations of reality. They exaggerate certain aspects to make them clearer. For example, no real structure can be magnified repeatedly an infinite number of times and still look the same. The principle of self-similarity is nonetheless realized approximately in nature: in coastlines and riverbeds, in cloud formations and trees, in the turbulent flow of liquids and in the hierarchical organization of living systems. It was Benoit B. Mandelbrot who opened our eyes to the *fractal geometry of nature.*
Actually, the processes which yield such structures have been studied for a long time in mathematics and physics. They are simple feed-back processes in which the same operation is carried out repeatedly, the output of one iteration being the input for the next one:

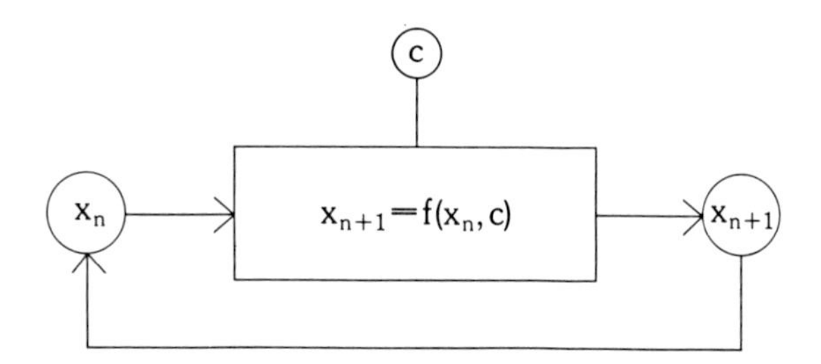


The only requirement here is a nonlinear relation between input and output, i.e. the *dynamical law* $x_{n+1}=f(x_n)$ must be more than a simple proportionality, $x_{n+1}=kx_n$. The schematic diagram indicates that the rule $x \mapsto f(x)$ will depend on a parameter *c*, whose influence will be discussed below.
If we start this kind of iterative process with an arbitrarily chosen $x_0$, it will generate a sequence of values $x_1, x_2, \ldots$, whose behavior over long periods of time is our interest. Will the sequence tend toward a particular limit value $X$ and come to rest there? Will it arrive at a cycle of values which is repeated over and over again? Or is the sequence erratic for all times, determined by the dynamic law and the initial value but nevertheless unpredictable?
Processes of this kind are found in all exact sciences. Indeed, the description of natural phenomena by differential equations, introduced by Sir Isaac Newton and Gottfried W. Leibniz some 300 years ago, is based on a feedback principle. The dynamical law determines the location and velocity of a particle at one time instant from their values at the preceding instant. The motion of a particle is understood as the unfolding of that law. It is not essential whether the process is discrete – that is, takes place in steps – or continuous. Physicists like to think in terms of infinitesimal time-steps: *natura non facit saltus.* Biologists, on the other hand, often prefer to look at the changes from year to year or from generation to generation. Obviously, both views are possible, and the circumstances stipulate which description is appropriate.

*The Scenario of Branching into Chaos*

Consider an example. The growth of a population over a period of years is usually described in terms of the growth rate, i.e. the yearly increase of the population relative to its total size. If that is constant over the years, the growth law is said to be linear, and the growth is called exponential. At a growth rate of 5 percent, for example, the population doubles every 14 years. These kinds of laws, however, apply for limited time spans. There are always limits to growth.
One of the first to notice this was P.F. Verhulst who 1845 formulated a law containing the growth limitation. He argued that a given niche can only maintain a certain maximal population *X* and that the growth rate should drop when the population size approaches X. Thus he was led to postulate a variable growth rate. The process thereby became *nonlinear,* with drastic consequences for its dynamic behavior.
It took more than a hundred years before all the ensuing complications were fully explored. For small growth rates, to be sure, nothing spectacular happens; the population size is simply regulated to stay at its optimal value, *X,* increasing when it is below, decreasing when it is above. As the rates exceed 200 percent, however, we are in for surprises.
Do we ever find growth rates this large in reality? Of course human populations do not grow this fast, but for certain insects it is not unusual. The important thing is that in the last 20 years Verhulst's law has found application to a much broader range of phenomena than he originally conceived. Edward N. Lorenz, a meteorologist at MIT, discovered in 1963 that precisely this law describes certain aspects of turbulent flow, particularly where the rate is high. Since then, theoretical work in laser physics, hydrodynamics, and the kinetics of chemical reactions has demonstrated the paradigmatic character of this law, and the scenarios predicted by it have been found in experiments.
But how does the Verhulst process behave when the rate becomes large? A detailed account is given in the Special Section 1.
Let us only mention the most important findings. At growth parameters slightly above 200%, the optimal size *X* cannot be reached any more. The vigorous growth of smaller populations invariably leads to an overshoot followed by a counterreaction in which the population decays to sizes well below the value X. After a while, a steady oscillation develops between two sizes, one large and one small (see Fig. 18).
When the growth parameter is increased beyond 245%, more complicated behavior ensues. The oscillations involve 4, then 8, then 16, different sizes, and so on, until for parameters larger than 257% *chaos sets in.*
What do we mean by chaos? In simple terms, the system has gone out of control. There is no way to predict its long time behavior. The irregular up and down in Fig. 20 persists and never turns into an orderly sequence of events. To appreciate the surprise that Lorenz felt at this discovery remember that there is no arbitrariness involved. The process is still described by Verhulst's law, the sequence is *determined* by its initial value – and yet, it *cannot be predicted* other than by letting it run.
This very subtle situation needs some further explanation. Saying the se-

quence is determined by the initial value presumes that the latter can be specified with infinite precision. This is only true “in principle”. Any real description of the initial size, its representation in a computer for instance, can only be given with finite precision. The process can be viewed as an unfolding of information: the longer we observe it the better we know, in retrospect, the exact value of the initial value.
The most exciting aspect of the Verhulst dynamics is not the chaos as such, however, but the *scenario by which the order turns into chaos.* It might seem hardly worthwhile to determine the exact growth parameters at which the bifurcations from period $2^n$ to period $2^{n+1}$ oscillations take place. Who cares? But pedantry has often been the godfather of important discoveries. Johannes Kepler would not have found the elliptical form of the planetary orbits had he not been disturbed by the tiny deviation in Mars' orbit of 8 minutes of arc from Ptolemaic theory. Friedrich Wilhelm Bessel would not have been able to determine the distance from the sun to the nearest fixed stars had he not learned the most exacting use of numbers and tables during his apprenticeship with a merchant in Bremen. Scientific work always depends on the most scrupulous attention to detail even when qualitative aspects are being clarified. There is no better tool for this than the computer, as anyone knows who has had to find errors in a program.

An exact analysis of the bifurcation points in the Verhulst process reveals a law of basic importance for the world of nonlinear phenomena. This concerns the lengths of the parameter intervals in which a particular period is stable. The intervals are shortened at each period doubling by a factor which approaches a *universal* value

$$\delta = 4.669\,201\,660\,910\ldots$$

as the period increases.

This number, whose initial decimal places were reported for the first time in 1977 by Großmann and Thomae, appears again and again for many other processes. It is as characteristic of the period doubling scenarios as the number $\pi$ is for the relation of the circumference to the diameter of the circle. This number is now called the “Feigenbaum number”. Mitchell Feigenbaum did calculations for many different processes on his desk computer at Los Alamos, and always arrived at the same factor. He discovered the universality of this number.
This discovery has spurred an enormous activity among scientists of many fields. A large number of experiments have been performed showing that the period doubling scenario actually occurs in many natural systems. Whether it is the onset of turbulence in fluid flow, nonlinear oscillations in chemical or electrical networks, or even the transition of the normal rhythm of the heart to life-threatening fibrillation, we simply cannot list all the groups in the United States, France, Germany, and elsewhere who have demonstrated that essential aspects of the dynamics of complex systems can be reduced to that behavior exemplified by the Verhulst equation.
The impact on theory was no less impressive. Mathematicians are still trying to fully understand that unexpected universality. But perhaps more impor-

tant, it has boosted a general hope that nonlinear phenomena may not be out of reach of systematic scientific classification.
The biologist Robert M. May was one of the first to recognize the importance of studying the Verhulst process. As early as 1976 he wrote:
*I would therefore urge that people be introduced to, say,* (the Verhulst equation) *early in their mathematical education. This equation can be studied phenomenologically by iterating it on a calculator, or even by hand. Its study does not involve as much conceptual sophistication as does elementary calculus. Such study would greatly enrich the student's intuition about nonlinear systems.*
*Not only in research, but also in the everyday world of politics and economics, we would all be better off if more people realised that simple nonlinear systems do not necessarily possess simple dynamical properties.*

## *Border Skirmishes: Chaos from Competition*

The significance of the bifurcation scenario has become basic to our understanding of nonlinear phenomena. The analysis of the Verhulst process made the idea of deterministic chaos an important topic and revealed some universal aspects of complex dynamical processes in an elementary way. This universality should not be misconstrued. There are, of course, other paths to chaos; indeed, other scenarios of equally general character have been discovered. The notion of universality in part reflects a tendency of physicists and mathematicians to borrow words that sound important. What is meant is that a certain behavior is *typical* and is more or less surprisingly found in a variety of systems.
It is clearly desirable to identify principles that show relations between individual scenarios. Benoit B. Mandelbrot achieved this in 1980 when he discovered the set that is today named in his honor. It is not just a bizarre figure which one finds pretty or ugly according to taste; the Mandelbrot set embodies a principle of the transition from order to chaos more general than the Feigenbaum universality. The aesthetic charm correlates with a fundamental meaning, as is often the case in mathematics.
Mandelbrot's ingenuity was to look at *complex* numbers instead of real numbers, to follow the process $x_0 \mapsto x_1 \mapsto x_2 \dots$ on a plane rather than on a line. The reader who is not familiar with complex numbers does not have to feel lost here; it is sufficient just to imagine that the rule $x_n \mapsto f(x_n)$ tells where the point $x_n$ is supposed to go to in the *plane* rather than on the line. The details of the rule are not essential because we will see that different rules may lead to the same Mandelbrot set. More important is that the transition from order to chaos is described from a more general point of view. The focus has shifted to the nature of *boundaries* between different regions. We can think of centers – *attractors* – which compete for influence on the plane: an initial point $x_0$ is driven by the process to one center or another, or it is on the boundary and cannot decide. If the parameter is changed, the regions belonging to the attractors change, and with them the boundaries. It can happen that the boundary falls to dust, and this decay is one of the most important scenarios.

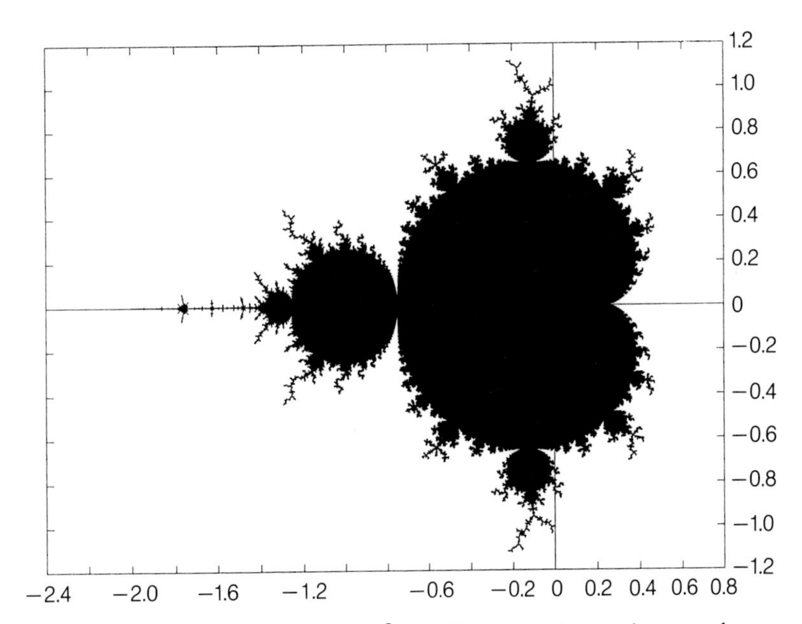


*Fig. 2. The Mandelbrot set for the process $x \mapsto x^2 + c$. The complex c-plane is shown with the window $-2.4 < \mathrm{Re}\, c < 0.8$, $-1.2 < \mathrm{Im}\, c < 1.2$. The figure reflects an ordering among the various types of boundaries at different values of the parameter c*

Mandelbrot's process is mathematically equivalent to the Verhulst process.

$$x_{n+1} = f(x_n) = x_n^2 + c$$

is the simple formula. Given a number $x_0$, take its square and add a constant $c$ to get $x_1$; then repeat to get $x_2$, $x_3$, and so on. Everyone can do that. But no one expected to find so much enigmatic beauty hidden in this iteration.
Let us begin with the simplest possible choice of the constant $c$, viz. $c = 0$. Then we just square the number at each iteration: $x_0 \mapsto x_0^2 \mapsto x_0^4 \mapsto x_0^8 \mapsto \ldots$.
There are three possibilities for the sequence, depending on $x_0$:

1. The numbers become smaller and smaller, their sequence approaches zero. We say that zero is an *attractor* for the process $x \mapsto x^2$. All points less than a distance of 1 from this attractor are drawn into it.
2. The numbers become larger and larger, tending towards infinity. We say that infinity is also an attractor for this process. All points farther than a distance of 1 from zero are drawn into it.
3. The points are at a distance of 1 from zero and stay there. Their sequence lies on the *boundary* between the two domains of attraction, in this case the unit circle around zero.

The situation is clear. Two zones of influence divide up the plane, and the boundary between them is simply a circle.
The surprise comes when we pick $c$ to be non-zero, say $c = -0.12375 + 0.56508i$. Here, too, the sequence $x_0 \mapsto x_1 \mapsto x_2 \mapsto \ldots$ has the choice between the three possibilities listed above, but the inner attractor (dot in Fig. 3) is no

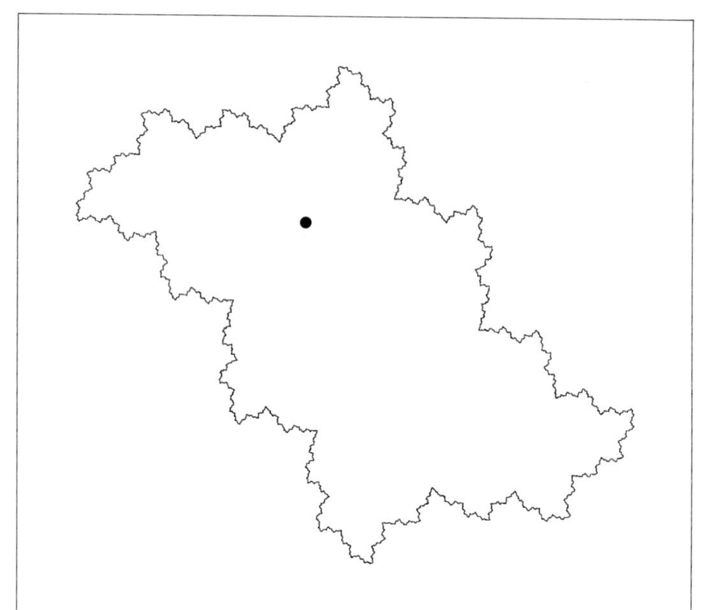

*Fig. 3. Basin of an attractive fixed point*

*Fig. 4. Basin of an attractive cycle of period 3*

longer zero and the boundary is no longer smooth. Figure 3 shows the crumpled course of the boundary. If we took a magnifying glass and looked more closely at the boundary, we would find that it looks exactly as crumpled as without the glass. This is what B. Mandelbrot calls the *fractal structure* of such a boundary. It reminds one of coast lines, of many natural boundaries which apparently become longer the finer the scale on which we measure them. One of the peculiar things about the boundary is its *self-similarity*. If we look at any one of the corners or bays, we notice that the same shape is found at another place in another size.

Boundaries of this kind have been known in mathematics as *Julia sets*. During the first world war, the French mathematicians Gaston Julia and Pierre Fatou studied their properties for the more general case of rational mappings of the complex plane (see the Special Section 2).

Their fascinating work remained largely unknown, even to most mathematicians, because without modern computer graphics it was almost impossible to communicate the subtle ideas. The self-similarity, for example, was well known to Julia and Fatou who stated that the entire boundary can be regenerated from an arbitrarily small piece of it, by a finite number of iterations of the formula $x \mapsto x^2 + c$ (compare 2.8). But how much easier is it to comprehend this feature by looking at pictures such as Figs. 3, 4, 6–15 than by thinking about the meaning of that statement.

Another general feature of Julia sets is that they carry incredibly complex dynamics. On the boundary the process is as chaotic as can be. The static Fig. 3 cannot show this, the color picture Map 20 gives a hint at best. The Julia set contains the unstable fixed point of the mapping $x \mapsto x^2 + c$ together with all its preimages; it contains an infinite number of unstable periodic sequences also

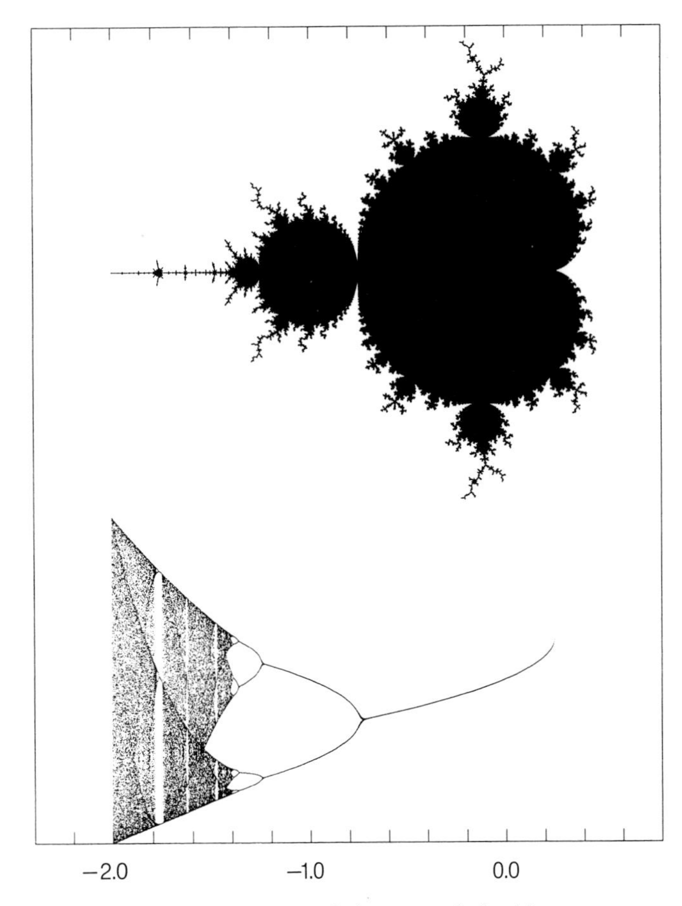


*Fig. 5. The relation between the Mandelbrot set and the period doubling scenario which takes place if c is varied as a real parameter. Bifurcations correspond to budding, periodic windows interrupting the chaotic veil correspond to small copies of the Mandelbrot set located on the main antenna*

with their preimages, and above all it contains chaotic sequences of points that never approach any kind of regularity. This can only be fully appreciated at the computer display, "in vivo". Or one can translate the picture into tones and follow the subjective effects of this "music". A fascinating effect is the so-called *intermittency* which occurs when the process comes into the neighborhood of a periodic point: the same theme recurs more or less frequently. When it finally breaks up the ear perceives a particular tension. The corresponding visual effect is not as strong.

If we go on to another choice of $c$, say $c = -0.12 + 0.74i$, we get Fig. 4. Here, the Julia set is no longer a single, deformed circle but consists of an infinite number of deformed circles which still constitute a connected set. The interior of this set is attracted not by just one fixed point but by a three-cycle, shown in the figure as three thick dots.

There is a principle that tells us which kind of Julia set a given choice of $c$ implies. This principle is the Mandelbrot set $M$. Fig. 2 shows it in black as part of the complex $c$-plane. A given complex number $c$ either belongs to the black

structure $M$ or it does not. The corresponding Julia sets of the process $x \mapsto x^2 + c$ are very different. They are connected structures if $c$ is from $M$, and broken into infinitely many pieces if $c$ lies outside. Therefore the boundary of $M$ is of particular interest. Imagine a path in the $c$-plane which begins in $M$ and terminates outside of $M$. As one varies $c$ along this path, the associated Julia sets will experience a most dramatic qualitative change as $c$ crosses the boundary of $M$: as if subject to an explosion, they will decompose into a cloud of infinitely many points. In this sense the boundary of $M$ locates some kind of a *mathematical phase transition* for the Julia sets of $x \mapsto x^2 + c$. Moreover, each of the various parts of $M$ represents a qualitative statement about the Julia sets for $c$-values taken from that part. The cardioid shaped main body, e.g., contains all values of $c$ for which the Julia set is a more or less deformed circle, surrounding the domain of attraction of a single fixed point (Fig. 3). Each bud on $M$ corresponds to a cyclical attractor of a particular period by a scheme which is well understood. The $c$-value of Fig. 4 is the center of the large bulb at the top of the main part of $M$. The three-cycle is created by the trifurcation of a fixed point when the parameter $c$ goes from the main part to the bud. The Verhulst period doubling scenario takes place on the real axis. Period 2 is stable in the large bud which includes $-1.25 < c < -0.75$ on the real axis and connects to the main body at the left side. The point $c = -2$ is the end point of the antenna of $M$ and corresponds to the value $r = 3$ in the Verhulst process (see Special Section 1). Figure 5 illustrates this connection and shows clearly how Mandelbrot's step into the complex plane gives a much more complete picture than real analysis does.
What does the Julia set look like when $c$ is a point of $M$ where a bud is connected to the main body? An example is shown in Fig. 6.

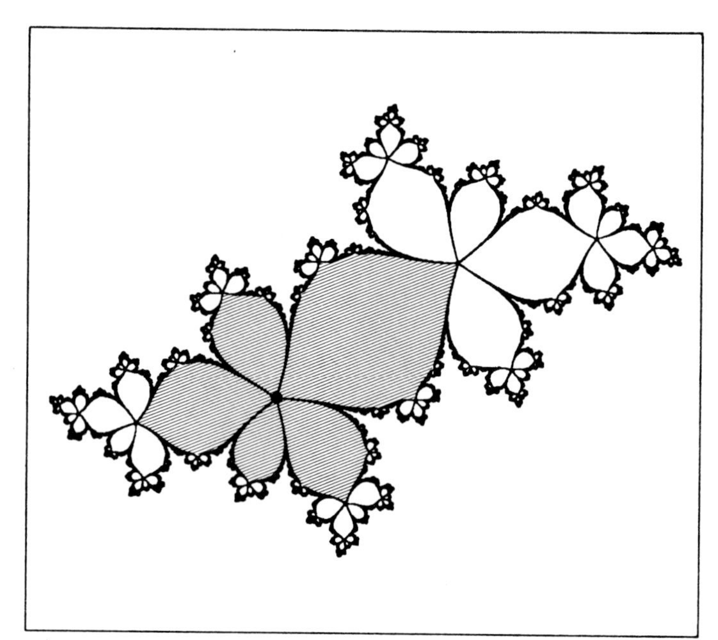

*Fig. 6. Parabolic basin around a fixed point*

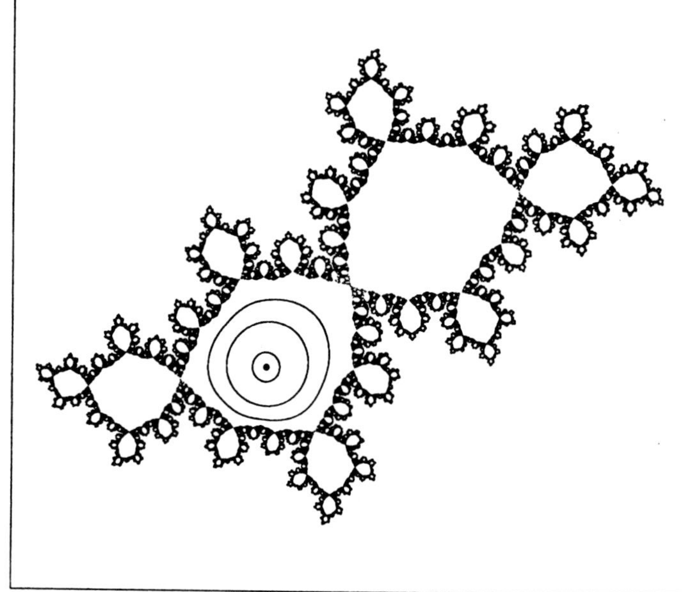

*Fig. 7. Siegel disk around a fixed point and pre-images*

The value $c = -0.481762 - 0.531657i$ corresponds to the germination point of a bud for stable 5-cycles. The 5 points of these cycles split off from the thick dot in Fig. 6 when $c$ moves into the bud. At the branch point, the Julia set reaches into the now marginally stable attractor. This is called the *parabolic* case of the dynamics. Figures 8 and 9 are further examples of this kind of Julia set.
Besides the germination points for buds, the main body of the Mandelbrot set has still other kinds of boundary points. When $c = -0.39054 - 0.58679i$ we get Fig. 7, where the fixed point is also marginally stable. Unlike in the parabolic case, the boundary does not extend to the fixed point, but the dynamics of the process does not reach it either. The circles shown around the fixed point are *invariant circles*, i. e. if one takes an initial point on one of the circles, all iterated points will also be on it. Inside the region bounded by the Julia set, the process runs in the following way: first the point springs from smaller, peripheral buds into larger ones until it comes into the disk containing the fixed point. This disk is called a *Siegel disk* after the German mathematician Carl Ludwig Siegel. When a point has arrived there, it simply rotates around the fixed point on its invariant circle.
With these four examples we have enumerated the typical cases where the boundary produced by the process $x \mapsto x^2 + c$ defines an interior region. In summary:

- If $c$ is in the interior of the main body of the Mandelbrot set, a fractally deformed circle surrounds one attractive fixed point (Fig. 3).
- If $c$ is in the interior of one of the buds, then the Julia set consists of infinitely many fractally deformed circles which surround the points of a periodic attractor and their pre-images (Fig. 4, 10).
- If $c$ is the germination point of a bud, we have a parabolic case: the boundary has tendrils that reach up to the marginally stable attractor (Fig. 6, 8, 9).
- If $c$ is any other boundary point of the cardioid or a bud (there are some technical conditions regarding the irrationality of the point), we have a Siegel disk (Fig. 7).

In a fundamental mathematical work in 1983, Dennis Sullivan showed that these four cases describe all the possible characteristic structures that a region contained by a Julia set can have except for one. A fifth possibility, the so-called *Herman ring* does not occur for the case $x \mapsto x^2 + c$; though proven to occur in other cases, it has never been observed. (See the Special Section 3.)
In Maps 18, 20, 24 colors have been used to exhibit an internal structure of the domains of attraction (and of the Siegel disk in Maps 22, 25). Map 20 shows the case in which there is one fixed point in addition to the attractor at infinity. The gradation of colors indicates how many iteration steps of $x \mapsto x^2 + c$ are necessary for a point to reach a selected small disk around the attractor. The same color means the same dynamic distance from the respective center of attraction. Thus, in the exterior region the coloring quantifies the escape toward infinity, while the interior is colored according to the

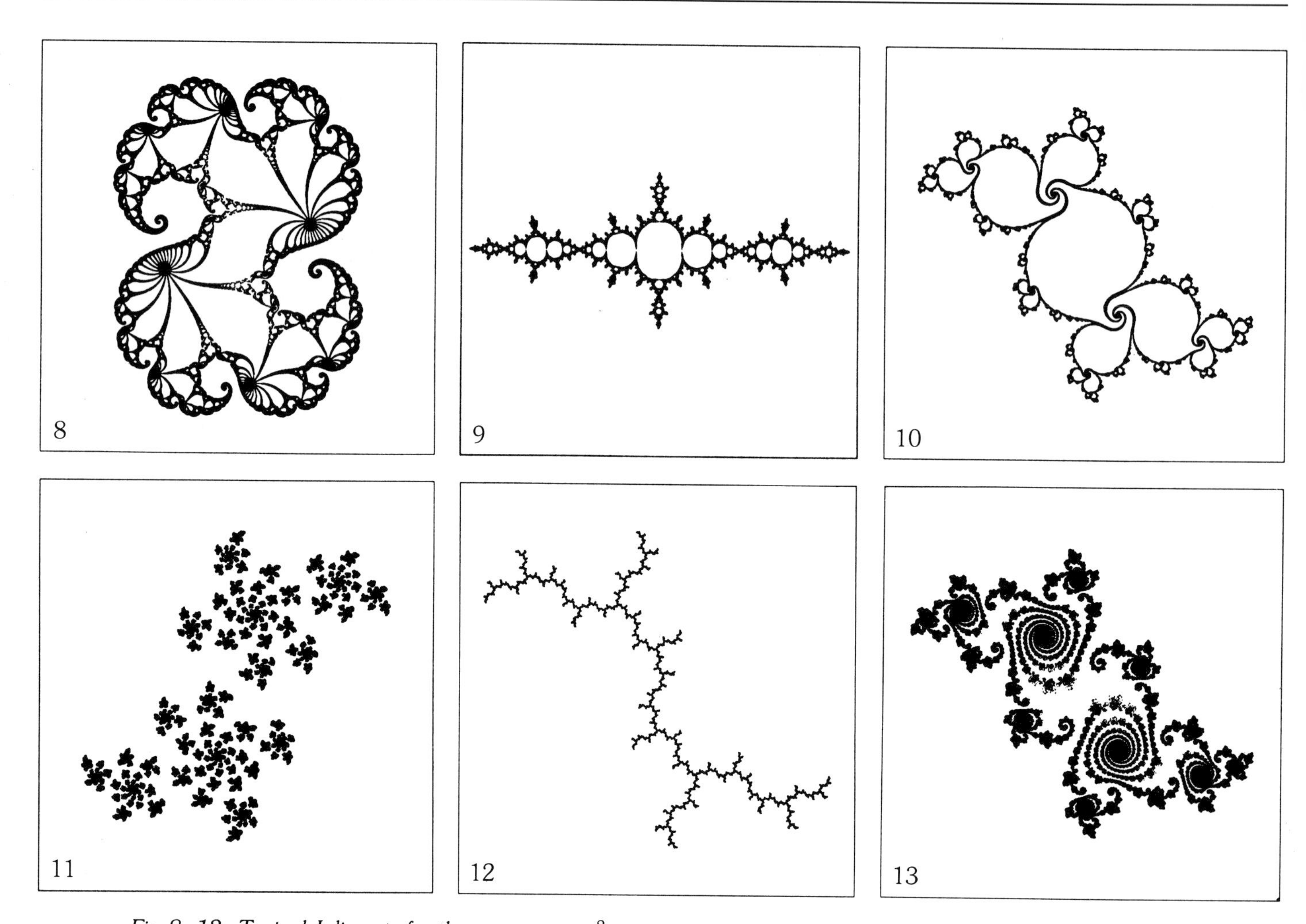


*Fig. 8–13. Typical Julia sets for the process $x \mapsto x^2 + c$*

*Fig. 8. Parabolic case; by a suitable arbitrarily small change of c, the marginally stable fixed point develops into an attractive cycle of period 20*
*Fig. 9. Parabolic case $c = -1.25$; $c > -1.25$: attractive cycle of period 2; $c < -1.25$: attractive cycle of period 4*
*Fig. 10. Connected Julia set (attractive cycle of period 3) shortly before it decays into a Cantor set (see Fig. 13)*
*Fig. 11. Fatou dust*
*Fig. 12. Dendrite, $c = i$*
*Fig. 13. Cantor set that develops from Fig. 10 by slight variation of the parameter c*

course of the bounded dynamics. Map 24 shows an attractive three-cycle, Map 18 a cycle of period 11. In the case of the Siegel disk (Map 22), the level sets run parallel to the invariant circles.
These examples are by no means an exhaustive list of all possible structures of Julia sets; there are other possible values for the parameter $c$. As the pictures show, the Mandelbrot set $M$ is surrounded by hair-like, more or less branched and crumpled antennae. If we pick $c$ on one of these antennae, we obtain a similarly shaped Julia set. Figure 12 shows the example $c = i$. Such

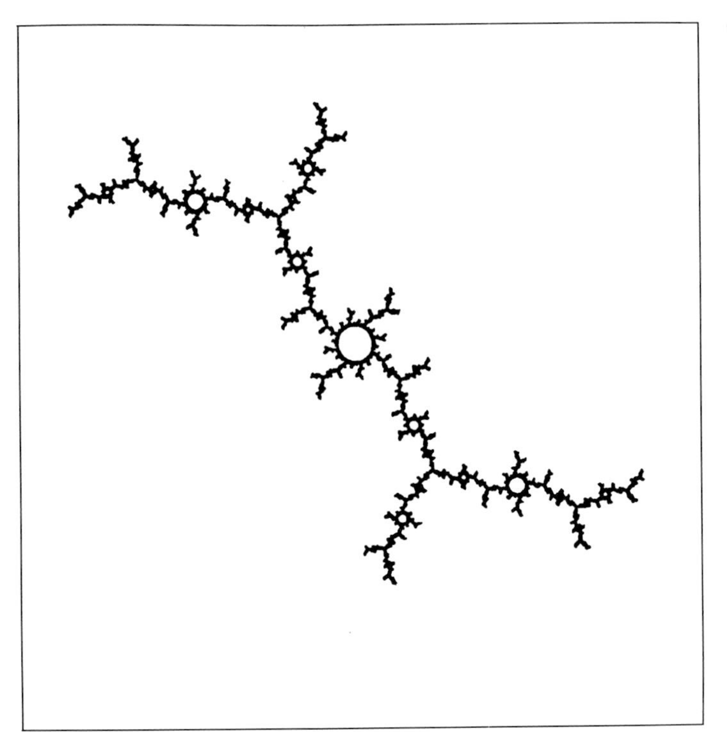

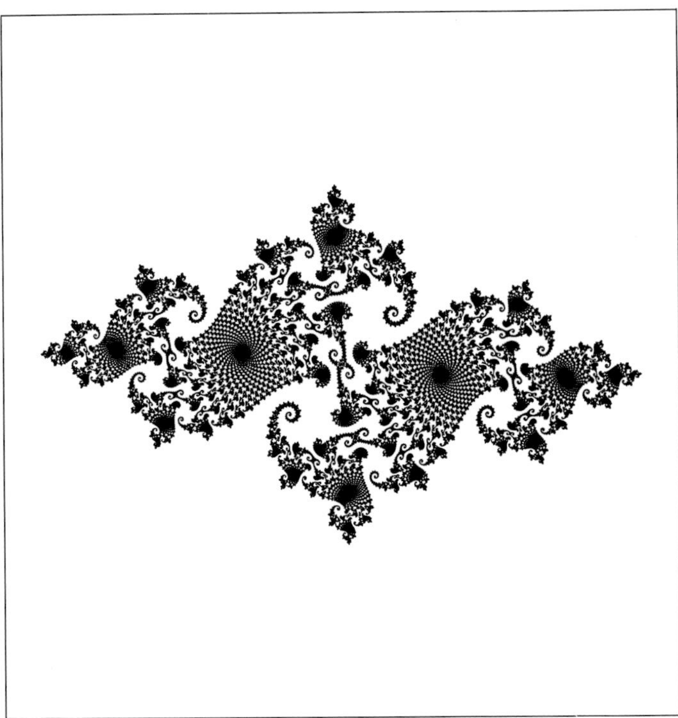

*Fig. 14. Dendrite with beads. Julia set for c-value from secondary Mandelbrot set*

*Fig. 15* (right). *Julia set for a c-value from seahorse valley*

dendrites have no interior; there is no attractor other than the one at infinity. The Julia set is now just the boundary of a single domain of attraction and contains those points that do not go to that attractor. As long as $c$ is in $M$ the Julia set has a connected structure. According to a theorem by Adrien Douady in Paris and John Hamal Hubbard in Cornell, the Mandelbrot set is also connected (see the Special Section 4).

One sees on close inspection that the antennae of $M$ carry many little copies of the larger $M$. They sit there as if on skewers, and further smaller ones sit between the larger ones, and so on and on forever. The windows with stable attractors which we saw in the chaotic region of the Verhulst process are cuts through these sprouts, as shown clearly in Fig. 5. If we pick $c$ in one of these miniature $M$'s, then the Julia set is a combination of a dendrite and the Julia set obtained from the corresponding value of $c$ in the main part of $M$, the latter copied infinitely many times and stuck onto the dendrites. Figure 14 shows an example for a value $c$ from the Mandelbrot set in Map 27.

Finally we can pick values of $c$ *outside* of $M$. As in the case of the purely dendritic structure, infinity is the only attractor, but now the Julia set has dissolved into a cloud of points called *Fatou dust.* This dust of points gets thinner and thinner the farther $c$ is from $M$. If $c$ is near the boundary of $M$, the dust makes fascinating figures as the examples in figures 11, 13, and 15 show, always fractal, self-similar, and carrying chaotic dynamics.

## *The Morphology of Complex Boundaries*

If the diversity of the Julia sets seems confusing, how much more confusing it would be without the Mandelbrot set! This guide through the world of parameters tells us what kind of boundary we should expect for a given value of $c$. This makes the boundary of $M$ particularly interesting, since it marks a change in the nature of the Julia sets. At the edge of the Mandelbrot set the Julia sets lose their connectedness, break up, and fall to dust.

A considerable number of our pictures are from this border zone. We discover a fantastic world there, whose richness of forms contrasts almost grotesquely with the simplicity of the formula $x \mapsto x^2 + c$. But is this not a familiar experience, that diversity flourishes particularly well at boundaries? Clear contours are the exception in the tension between opposing principles. Every larger conflict reveals a thousand smaller ones. In this way, the general course of the border corresponds to similar structures at finer and finer scales.

The qualitative jump at the boundary of the Mandelbrot set influences the region near this boundary. A simple black and white coloration will not show this (if black, e.g., is taken for connected, white for burst Julia sets). The complex dynamical structure of the border region can only be shown adequately with the help of color graphics. Even the 256 colors available for our pictures can merely allude to the dynamics. It takes interactive experimentation at the graphics terminal to grasp its true complexity.

How do we arrive at the colors for the neighborhood of $M$? Imagine it to be made of metal, electrically charged. The surface then has a constant electrical potential of, say, 1000 volts. This falls off to zero in the region surrounding the conductor, and we trace out lines of constant voltage, the so-called equipotential lines (Fig. 16). The 1 volt line, for example, is so far away from the con-

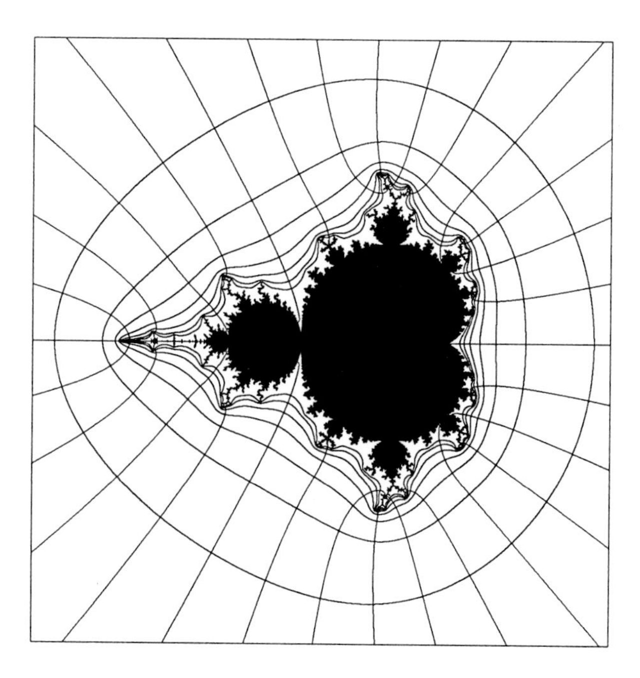

*Fig. 16. Charged Mandelbrot set with equipotential and field lines*

ductor that it is nearly circular, because from that distance $M$ almost looks like a point charge. The 900 volt line, in contrast, gives some clue as to the form of $M$, and the 999 volt line follows the contours fairly closely. Our coloration follows these lines. The points between two such lines have one color. The various colors give a contour map of the electrostatic potential between the boundary of $M$ and infinity.

Now, what have these equipotential lines to do with the dynamics of the process $x \mapsto x^2 + c$? In 1983, A. Douady and J. Hubbard proved the amazing mathematical fact that the equipotential lines show precisely the dynamics of the *critical point* $x = 0$. The equipotential lines are also lines of equal escape time towards infinity for the initial point $x_0 = 0$ (see Special Section 4).

This relationship between the electrostatic picture and the dynamics suggests a simple way to determine contour lines on the computer. Those values of $c$ for which the critical point needs a particular number of iterations to get outside a circle of radius 1000 exactly fill a band between two equipotential lines. As we approach the boundary of $M$, the number of iterations becomes larger. The dynamics is trapped for a longer and longer time in the twisted paths of Julia's border region.

The color Maps 26–54 and 99–101 show enlarged sections from the border of $M$. (The small black and white figures serve to identify the location of the various Maps along $M$.) It is hard to believe the multitude of structures contained in the formula $x \mapsto x^2 + c$. Can there be a more striking demonstration of the enormous complexity hidden in the simplest laws?

Let us now contemplate the pictures and note some observations. It is apparent that each locus on $M$ defines motifs. If we wander along the border of $M$, we notice a gradual metamorphosis of these motifs, for instance starting with the cross in Map 30 and accumulating more und more twisted arms until the sea-horse of Map 36 emerges. At any given place, the motif is taken through an infinite number of variations. This is shown in the enlargement series in the "sea-horse valley", Maps 34–50, which up to a magnification of one million shows ever new constellations of "tail" and "eye" of the sea-horse.

Another very noticeable feature is the similarity in the structure of a detail of the Mandelbrot set to the form of the corresponding Julia set. The Julia set in Fig. 15 belongs to a value of $c$ in the vicinity of the tail of the sea-horse (Map 42). The qualitative similarity of the forms is amazing. It goes so far that the number of spiral arms which issue from the eyes is 29 in each case.

There is one constant in the diversity of motifs in this morphology of Julia sets: the Mandelbrot set itself, which appears again and again in different sizes but always in the same form. One is reminded of the genetic organization in higher organisms: each cell contains the complete genome, the totality of all forms of expression, but at any point in the organism only a small selection is actually expressed.

We mentioned before that $M$ has been shown to be a connected set; no part of $M$ is detached from the main body, but everything is held together by extremely thin lines. Maps 58–60 intrigue us most in this respect. The minute detail at the border of $M$, shown here in three different colors, gives an idea of the bewildering system of bridges necessary to secure the connectedness.

The viewer is invited to make his own associations with these pictures. We apologize if our own interpretations seem too fanciful. Naturally, we are biased toward a product which as a scientific result took many hours of computer time and as a color composition required no small effort on our part. We do believe however, that our fascination comes primarily from the basic issues, from the fantastic phenomenology of these complex boundaries so inviting to aesthetic involvement. We have to admit that some things about the pictures are not natural. The infinite microscopic depth to which the self-similarity seems to reach is a mathematical construct that does not exist in the real world. Physical objects are seldom self-similar over more than four orders of magnitude. In biology, a new principle of self-organization takes over usually after about 2 orders of magnitude (macromolecules have a diameter of about 100 atoms, simple cells a diameter of about 100 macromolecules, etc.). Therefore, the process $x \mapsto x^2 + c$ is not a proper description of the real world. But we did not assert that it was! Every law has its range of validity which must be sounded out. The range of validity of *linear* laws is no longer sufficient, at least in physics, so it is now necessary to find out how nonlinear laws can help us understand the universe. In this way the study of the quadratic law $x \mapsto x^2 + c$ has a very fundamental meaning. Mandelbrot's discovery of the universal figure $M$ is without doubt an event of great consequence for the theory of dynamical systems.

*Newton's Complicated Boundaries*

Sir Isaac Newton laid the foundations of classical mechanics, of optics, and of calculus. But in addition of those basics of the natural sciences he established a number of less well known methods that have proven useful up to the present day. In fact, the so called Newton algorithm for finding the roots of an equation $f(x) = 0$ is being used more than ever before, now that computing machines can apply the method with so much more ease and accuracy than humans can. A number of the pictures in this book celebrate this very method.

Newton's algorithm is an ingenious trick. It transforms the problem of solving the equation $f(x) = 0$ into a dynamical process where the solutions compete for initial guesses. One does not need to know the correct solution to begin. When one starts with any odd guess, Newton's recipe leads one closer to one of the solutions. They act as the centers of an attractive field of force (one of Newton's favorite themes!).

How far does the attractive influence of the various centers extend, and what does the boundary between them look like? Lord Arthur Cayley in 1879 was the first to address this question seriously; and he had to give up eventually, because he found the answer too complicated. The details of Cayley's analysis and subsequent developments are presented in the Special Section 6. Suffice it here to mention that the problem encountered by Lord Cayley was the starting point from which Julia and Fatou developed their magnificent iteration theory for rational functions in the complex plane.

We have not mentioned one trait of the Julia sets, even though it lends particular charm to such examples as the magnetism picture of Maps 6 and 10. No matter how many attractors are distributed over the plane, each point of the Julia set simultaneously touches all their domains of attraction. For the case of three attractors, each point of the border is a place where three domains meet! That sounds impossible, but the "planet" in Map 75 shows how it works. It displays in colors yellow, blue, and grey the domains of influence of the roots of a polynomial equation, as defined by Newton's algorithm. Wherever two regions are about to form a boundary (yellow and blue, for example), the third region (grey) establishes a chain of outposts. In order that these outposts do not form bilateral borders with their neighbors, they in turn are surrounded by chains of islands in a structure which is repeated down to infinitely small dimensions. The little moon in Map 75 has been inserted to show the rear side of the planet. Map 76 shows the same planet in different colors. Here the complex boundary structure (i.e. the Julia set) is enhanced in the central region by a white shimmer.

What may seem almost impossible as a boundary between three "countries" can be extended without any mathematical difficulty to situations with 4, 5, 6, ... competing domains. The boundary is made up entirely of points where 4, 5, 6, ... countries meet. Map 61 shows the polar cap of a planet whose four domains (red, green, blue, and yellow) form a boundary of four-corner-points. Map 62 shows the same region in different colors, while Map 63 is another rendering of the same structure. The grey moons, which have been added arbitrarily, carry the same structure as in Map 75.

Maps 90–98 show step by step the result of another kind of animation. Map 90 repeats the structure of Maps 75 and 76, with boundaries made up of 3-corner-points. Map 98, on the other hand, is close to being another version of Maps 61–63 where the Julia set consists of 4-corner-points. The sequence shows how our artificial planet grows new continents, by changing smoothly from the 3-domain structure to the case of 4 domains. The percentages indicate how far the animation has proceeded. The boundaries seem to be undergoing certain distinct crises; one is easily reminded of fault lines. Map 89 shows the particularly interesting intermediate structure of Map 94 in different colors.

Newton's method is not necessarily confined to problems in the complex plane. The complex numbers have the advantage that there is a well-developed mathematical discipline to guide us, namely, the iteration theory of Julia and Fatou. There is no corresponding theory for problems with real functions. And yet some of our most appealing pictures have been obtained in this context. For details of their mathematical background we refer the interested reader to Special Sections 7 and 8. It is obvious that Maps 17, 19, 21, 23, 55–57, 67–74, 87, 88 are of a different nature than the Maps considered thus far. Processes in the complex plane seem to produce almost baroque patterns, whereas real iterations tend to generate more modern shapes. The reader may amuse himself by guessing what kind of iteration he is looking at. We do not know precisely what causes these different impressions; but we must also admit that the purely mathematical features of the latter pictures are far from being well understood.

*Science and/or Art*

As we entertained the thought of a public exhibition of "pictures from the theory of dynamical systems" in the summer of 1983, we thought that the aesthetic appeal of the pictures themselves would be sufficient. How naive we were, and how we underestimated our public! What had been quite simply fun in the context of our scientific work suddenly became the topic of very serious discussions. The viewers demanded an explanation of that context and wanted to know its importance. We suddenly incurred an obligation to clarify the content symbolized by our pictures.

Some of our esteemed colleagues worried that something might be expanding into art that had no scientific certification. Fritz Meckseper, on the other hand, a well known artist from Worpswede whose lifelong pursuit had been to reduce his perception of the world to symbolic representations, asked what we needed the pictures for if we had the formula $x \mapsto x^2 + c$. An art critic from "Die Zeit" denied that our pictures could qualify as art, on the grounds that they were lacking an element of choice or of free expression. A scientific friend who had very consciously turned to poetry and painting as means of expressing his thoughtful emotions, felt our work lacked an ingredient of human concern.

We hesitate to comment on these responses. *"Art is a lie that lets us recognize the truth"*, said Pablo Picasso. It may not be a deep truth to assert that our world is nonlinear and complex; everyday experience has never taught us otherwise. Yet physics and mathematics, and other sciences following them, have successfully managed to ignore the obvious. Focussing on simple problems that they could solve, these sciences had a strong impact on technology and thereby drastically changed the surface of our planet. It is now being felt, however, that more is needed than knowledge of the linear phenomena. The concern is growing, almost simultaneously in many different fields, that far too little is known about the consequences of nonlinear laws. It came as a surprise even to physicists that there is chaos in their simplest equations. Our pictures express an optimistic attitude in this respect.

Mindboggling at first glance, they also show that the complex is accessible to systematic study, that even the chaos has its rules. The regularities of the Mandelbrot set encourage our hope that more characteristic scenarios will be identified in the nonlinear world. This hope rests on the power of computer experimentation which has so quickly become one of the main sources of insight and of inspiration.

But every tool requires a creative mind that puts it to good use. It would be unfair to discredit our pictures as being purely the result of machine work. They are not. There is plenty of freedom involved in their generation, both in an objective and in a subjective sense. As scientists, we choose specific questions on which we let the computer do its powerful work. When the machine has done its job we are left with a wealth of objective information that cannot be digested as such. Choices must be made. There are many ways to render that information accessible to further meaningful processing. In the assignment of colors, for instance, there is the very subjective choice of balance between integration and differentiation: some information is discarded by the use of the

same color for different points; other features are being enhanced by carefully adjusting their display to our aesthetic intuition.
In two years of trying to present our work to an interested public of widely different backgrounds, we have come to believe that artistic activity can also be scientifically fruitful. Or are all the many vows of mathematicians and theoretical physicists to the aesthetic component ot their science purely platonic? The American mathematicians Philip J. Davis and Reuben Hersh wrote:

*Blindness to the aesthetic element in mathematics is widespread and can account for a feeling that mathematics is dry as dust, as exciting as a telephone book, as remote as the laws of infangthief of fifteenth century Scotland. Contrariwise, appreciation of this element makes the subject live in a wonderful manner and burn as no other creation of the human mind seems to do.*

How could the aesthetic element come alive other than by being integrated into the search for mathematical and scientific knowledge?
Many reactions to our previous exhibitions have reinforced our conviction that if art and science would approach each other more closely, there might be great mutual benefit. The chance to do this is not to be understood just as an infatuation with something new and different, but considered quite realistically in the forms of the "new media", above all the computer. The computer is no longer the exclusive domain of scientists and engineers; a young generation of PC-acrobats is growing up, who will certainly develop their own artistic ambitions. It is unclear now where this development will go, unclear perhaps in the sense of complex dynamics: well determined but unpredictable, seething at its turning point like Faust at his rejuvenation in the witches cave:

*Not Art and Science serve, alone;*
*Patience must in the work be shown.*
*A quiet spirit plods and plods at length;*
*Nothing but time can give the brew its strength.*
*And all, belonging thereunto,*
*Is rare and strange, howe'er you take it.*
*The Devil taught the thing, 't is true,*
*And yet the Devil cannot make it.*

*J. W. von Goethe*

# SPECIAL SECTIONS

## 1 Verhulst Dynamics

### A Population Growth Model

Let $x_0$ be the initial population size and $x_n$ its value after $n$ years. The growth rate $R$ is the relative increase per year,

$$R = (x_{n+1} - x_n)/x_n.$$

If this is a constant $r$, the dynamical law is

$$(1.1) \quad x_{n+1} = f(x_n) = (1+r)x_n.$$

After $n$ years, the population size is $x_n = (1+r)^n x_0$. To limit this exponential growth, Verhulst assumed the rate $R$ to vary with population size. Arguing that a given niche can only sustain a certain size $X$ (which we may arbitrarily set equal to 1) he postulated the size-dependent rate $R$ to be proportional to $1-x_n$, i.e. we may set $R = r(1-x_n)$; the constant $r>0$ will be called the growth parameter. Hence, if $x_n < 1$ the population may still increase until it stops growing when $x_n = 1$. The dynamical law then takes the form

$$(1.2) \quad x_{n+1} = f(x_n) = (1+r)x_n - rx_n^2.$$

There are two choices for $x_0$ such that the population does not change its size: $x_0 = 0$ and $x_0 = 1$. If $x_0 = 0$ there is nothing to start with, so there cannot be any growth. However, if by chance there is a little bit, $0 < x_0 \ll 1$, then for $r>0$ there will be more the next year: $x_1 \approx x_0 + rx_0$. Thus the steady state $x_0 = 0$ is unstable. The sequence $x_0, x_1, x_2, \ldots$ will be growing until, hopefully, it reaches size 1. To determine whether $x_0 = 1$ is stable, we analyze how small deviations $\delta_n = x_n - 1$ develop in time. Linearizing (1.2) we find

$$(1.3) \quad \delta_{n+1} \approx (1-r)\delta_n$$

which shows that $\delta_{n+1}$ is smaller in magnitude than $\delta_n$, provided $0<r<2$. The case $r = 1.8$ is illustrated in Fig. 17 where the initial value is $x_0 = 0.1$. The value of $x$ increases at first because it is well below 1. At the third step it shoots somewhat above this level. Since according to (1.3), $\delta_{n+1} \approx -0.8\,\delta_n$, the deviations decrease in magnitude and the process settles down to the desired final state $x = 1$. For $r>2$, however, (1.3) predicts an increase of the deviations $\delta_n$ and we conclude that the steady state $x = 1$ has become unstable.
To proceed further we perform the experiment shown in Fig. 18. It shows that for $r = 2.3$ the process eventually becomes a periodic oscillation between two levels. This suggests to consider the first iterate of (1.2), $x_{n+2} = f(f(x_n)) = f^2(x_n)$, and to analyze the stability of the fixed points of $f^2$. We find that they are stable as long as $r < \sqrt{6} = 2.449$.

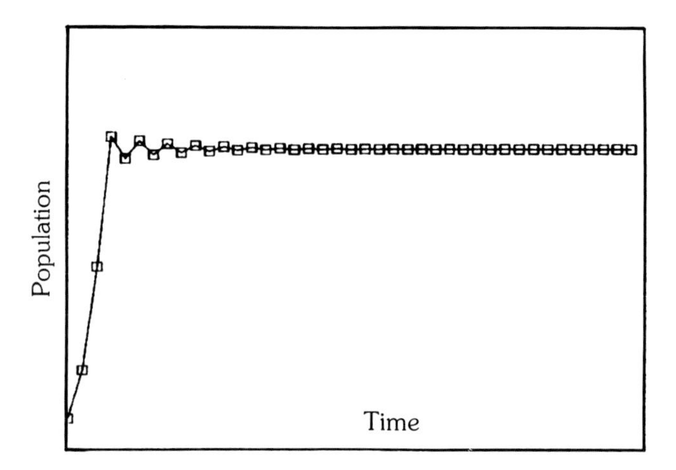


Fig. 17. $r=1.8$

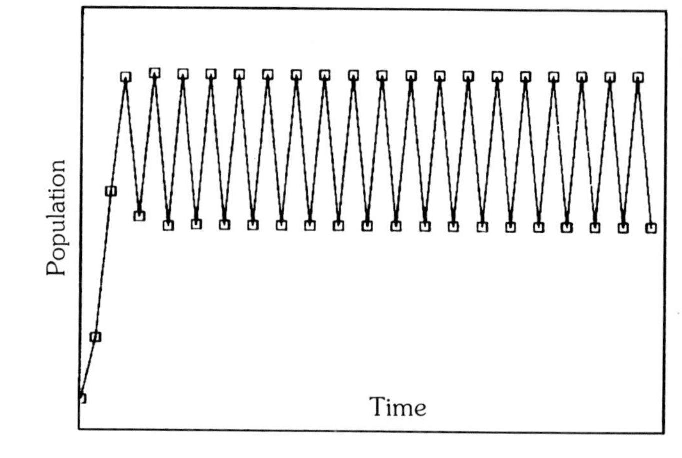


Fig. 18. $r=2.3$

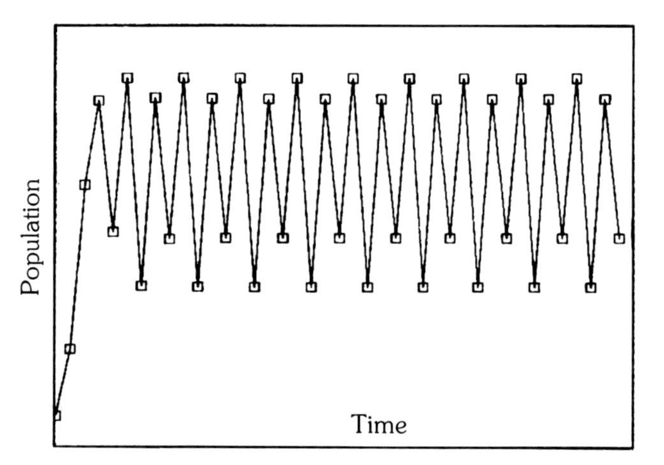


Fig. 19. $r=2.5$

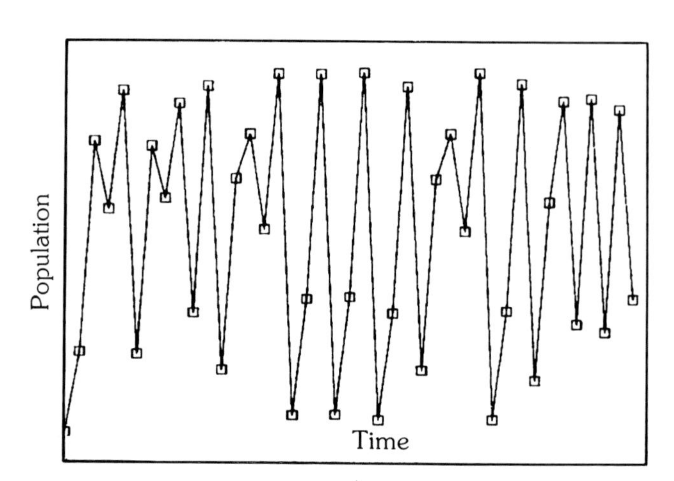


Fig. 20. $r=3$

*Transition from Order into Chaos*

The analysis becomes more and more difficult as $r$ increases. For $r=2.5$, Fig. 19 suggests that the process approaches a stable oscillation of period 4, and further experimentation reveals period doubling at ever more closely spaced values of $r$. Finally, at $r=2.570$, the process is no longer periodic at all. It jumps around incessantly among an infinite number of values in a way which is actually deterministic but cannot be predicted over long periods of time. The term *chaotic* has become customary for this behavior. As an example, the sequence we obtain for $r=3.0$ and $x_0=0.1$ is shown in Fig. 20.
If $r_n$ is the value of the growth parameter at the $n$-th bifurcation (where period $2^n$ becomes unstable and period $2^{n+1}$ gains stability), then the ratio of the lengths of successive intervals,

$$\delta_n=\frac{r_n-r_{n-1}}{r_{n+1}-r_n},$$

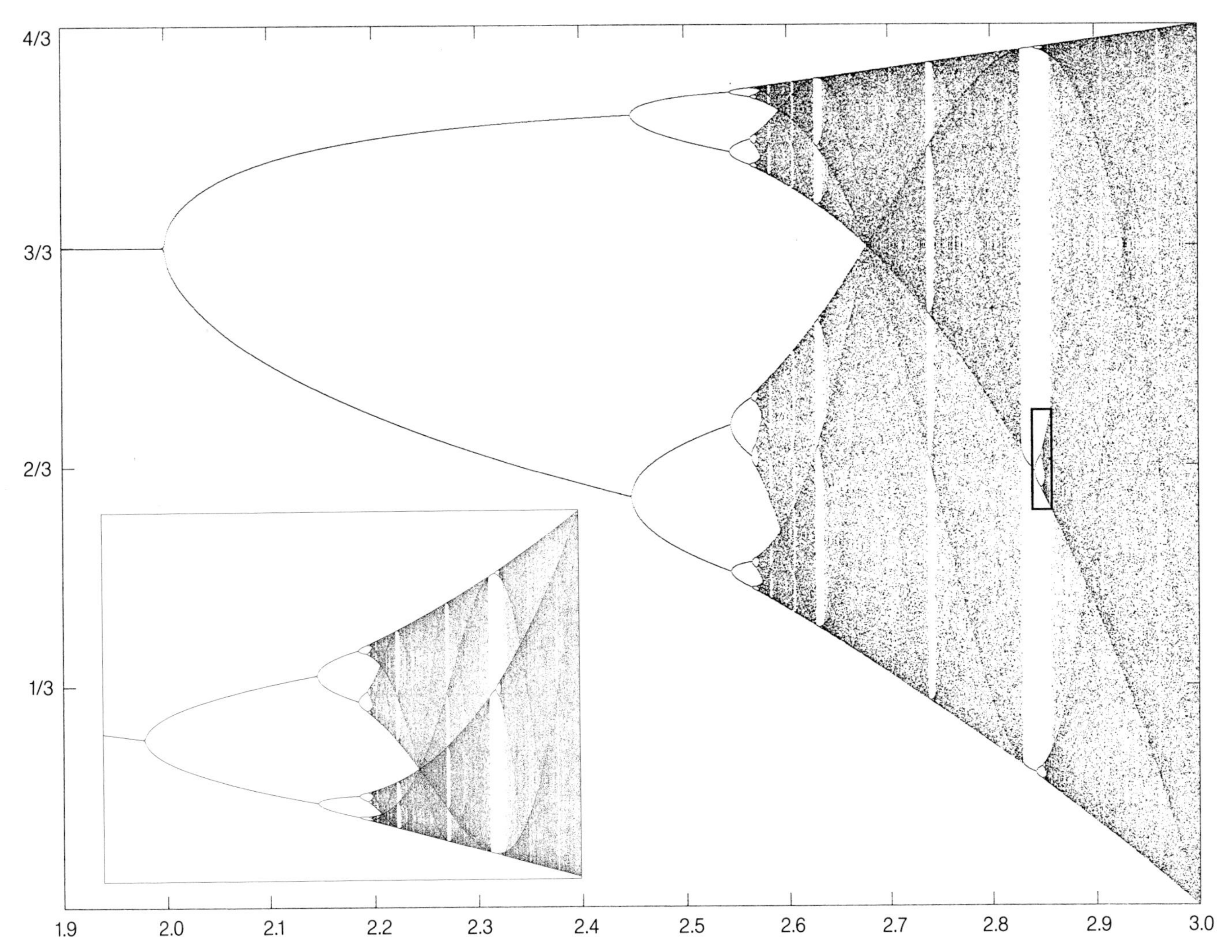


*Fig. 21. The period doubling scenario of the Verhulst process (1.2). The growth parameter r varies along the abscissa (1.9 < r < 3). For each value of r, 120 iterations of x are plotted* after *a transient period of* 5000 *iterations. The insert is a blow-up of the window indicated; the magnification is higher in the r-direction than in the x-direction*

has been found (by S. Großmann and S. Thomae, as well as by M. Feigenbaum) to approach the value

(1.4) $\delta_n \rightarrow \delta = 4.669\ldots \quad \text{as } n \rightarrow \infty.$

Feigenbaum showed that the *same number* $\delta$ appears in a variety of processes other than the Verhulst process, and is in fact a universal characteristic of the period doubling scenario in many one-dimensional processes.

We would of course like to have an overview of the various types of behavior of Verhulst processes. This is provided by the *bifurcation diagram* of Fig. 21 which shows the $r$-dependence of the asymptotic dynamics. For a given value of $r$, 5000 iterations were performed "in the dark" to let the process settle on its *attractor* (i.e. the asymptotic behavior after transient aspects have decayed away), and then 120 iterations were plotted to show the nature of that attractor. It consists of one point for $r<2$, of 2 points for $2<r<\sqrt{6}$, then of 4, 8, 16, ... points until in the chaotic regions the points of the attractor fill entire bands.

The cascade of bifurcations which E.N.Lorenz observed below the chaos-point $r=2.570$ has a corresponding structure above this point. This was first pointed out by Siegfried Großmann and Stefan Thomae at the University of Marburg in W.Germany: Near $r=3.0$ there is only *one* chaotic band which splits at $r=2.679$ into *two* bands, at $r=2.593$ into *four*, then into 8, 16, 32, and so on, until again at $r=2.570$ this doubling has taken place an infinite number of times.

Figure 21 contains a series of other bifurcation "trees" which are also characterized by the number $\delta$. In the midst of the chaotic region we see "windows" in which the attractor consists again of distinct points. For example, at $r=2.8284$ there appears a stable period 3 which doubles to 6, 12, 24, . . ., and disappears into chaos at $r=2.8495$.

For more detailed information we recommend the books by R.L.Devaney [De1] and H.G.Schuster [Sch].

# 2 Julia Sets and Their Computergraphical Generation

This is a collection of some fundamental classical results from the work of Gaston Julia (1893–1978) and Pierre Fatou (1878–1929). For a more detailed review we recommend [Bl].
As usual, we denote by $\mathbb{C}$ the Gaussian plane of complex numbers, and by $\overline{\mathbb{C}}$ the Riemann sphere $\mathbb{C} \cup \{\infty\}$. Let $R$ be a rational function, i.e.

$$R(x) = P(x)/Q(x),\ x \in \overline{\mathbb{C}},$$

where $P$ and $Q$ are polynomials without common divisors. The degree of $R$, $\deg R = \max\{\deg P, \deg Q\}$, is assumed larger than 1. The degree is also, in general, the number of preimages of a point $x$, $R^{-1}(x) = \{y \in \overline{\mathbb{C}}: R(y) = x\}$.
In a sense, the *Julia set* $J_R$ is a set of exceptional points for the iteration of $R$: $R^n(x) = R(\ldots(R(R(x)))\ldots)$ $n$ times, $n = 1, 2, 3, \ldots$. The complement of $J_R$ is called *Fatou set*, $F_R = \overline{\mathbb{C}} \backslash J_R$. The classical definition of a Julia set is, however, not very appealing to our intuition. We therefore choose another one which is easier to grasp. It requires that we first deal with *periodic orbits*. For every $x_0 \in \overline{\mathbb{C}}$, the recursion $x_{n+1} = R(x_n)$, $n = 0, 1, 2, \ldots$ defines a sequence of points. This sequence is called the *forward orbit* of $x_0$, and denoted by $Or^+(x_0)$. The definition of *inverse orbits* may seem to create problems since the inverse mapping $R^{-1}$ is multivalued. But collecting all preimages, we set

$$Or^-(x_0) = \{x \in \overline{\mathbb{C}}: R^k(x) = x_0 \quad \text{for} \quad k = 0, 1, 2, \ldots\}.$$

If $x_n = x_0$ for some $n$ in $Or^+(x_0)$ we say that $x_0$ is a *periodic point*, and $Or^+(x_0)$ is called a *periodic* orbit or *cycle*. We sometimes write $\gamma = \{x_0, R(x_0), \ldots, R^{n-1}(x_0)\}$. If $n$ is the smallest integer with that property, then $n$ is the *period* of the orbit.
If $n = 1$ we have that $R(x_0) = x_0$, i.e. $x_0$ is a *fixed point* of $R$. Obviously, $x_0$ is a fixed point of $R^n$, if $x_0$ is a periodic point of period $n$. (Iterates of $R$ should not be confused with powers of $R$, i.e. $R^n(x) = R \circ \ldots \circ R(x)$ is different from $(R(x))^n$.)
To characterize the stability of a periodic point $x_0$ of period $n$ we have to compute derivatives. The complex number $\lambda = (R^n)'\,(x_0)$ $\left(' = \frac{d}{dx}\right)$ is called the *eigenvalue* of $x_0$. Using the chain rule of differentiation we see that this number is the same for each point in a cycle. A periodic point $x_0$ is called

| | | | |
|---|---|---|---|
| *superattractive* | $\Leftrightarrow \lambda = 0$ | *indifferent* | $\Leftrightarrow \lvert\lambda\rvert = 1$ |
| *attractive* | $\Leftrightarrow 0 < \lvert\lambda\rvert < 1$ | *repelling* | $\Leftrightarrow \lvert\lambda\rvert > 1$ |

We can now characterize the Julia set $J_R$ of a rational function $R$: Let $P$ be the set of all repelling periodic points of $R$. Then

(2.1) $P$ is dense in $J_R$.

I.e. each point in $J_R$ is the limit of a sequence of points from $P$. If $x_0$ is an attractive fixed point we consider its *basin of attraction*

$$A(x_0)=\{x\in\overline{\mathbb{C}}: R^k(x)\to x_0 \text{ as } k\to\infty\};$$

$A(x_0)$ collects all points $x$ whose forward orbits $Or^+(x)$ approach $x_0$. This set includes, of course, the inverse orbit of $x_0$, $Or^-(x_0)$. If $\gamma$ is an attractive cycle of period $n$, then each of the fixed points $R^i(x_0)$, $i=0,\ldots,n-1$, of $R^n$ have their basins and $A(\gamma)$ is simply the union of these basins.
We now list a first set of fundamental results about $J_R$ from [Ju], [Fa]:

*Fundamental Properties of Julia Sets*

(2.2) $J_R\neq\emptyset$ and contains more than countably many points.

(2.3) The Julia sets of $R$ and $R^k$, $k=1,2,\ldots$, are identical.

(2.4) $R(J_R)=J_R=R^{-1}(J_R)$.

(2.5) For any $x\in J_R$ the inverse orbit $Or^-(x)$ is dense in $J_R$.

(2.6) If $\gamma$ is an attractive cycle of $R$, then $A(\gamma)\subset F_R=\overline{\mathbb{C}}\backslash J_R$ and $\partial A(\gamma)=J_R$.

(Here $\partial A(\gamma)$ denotes the boundary of $A(\gamma)$, i.e. $x\in\partial A(\gamma)$ provided $x\notin A(\gamma)$ but $x$ is an accumulation point of a sequence with elements from $A(\gamma)$.) Figures 3, 4, 10 and Maps 3–10, 18, 20, 22, 24, 25, 61–66, 75–78, 89–98 are examples of Julia sets bounding two, three or even four different basins of attraction of attractive fixed points.
Further results on Julia sets are:

(2.7) If the Julia set has interior points (i.e. there are points $\bar{x}\in J_R$ such that for some $\varepsilon>0$ $\{x: |x-\bar{x}|<\varepsilon\}\subset J_R$) then $J_R=\overline{\mathbb{C}}$.

This situation appears to be rare, but the mapping $R(x)=((x-2)/x)^2$ is an example. (See Special Section 3.)

(2.8) If $\bar{x}\in J_R$, $\varepsilon>0$, and $J^*=\{x\in J_R: |x-\bar{x}|<\varepsilon\}$, then there is an integer $n$ such that $R^n(J^*)=J_R$.

A number of comments seem to be in order. Property (2.2) implies that every rational map has a considerable repertoire of repelling periodic points. According to (2.4) the Julia set is invariant under $R$, and the dynamics on $J_R$ is chaotic in some sense as a result of (2.1). Property (2.5) suggests a numerical way to generate pictures of $J_R$. Unfortunately the inverse orbit of a point $\bar{x}\in J_R$ usually does not distribute uniformly over the Julia set. (See Fig. 27 for the distribution of $Or^-(\bar{x})$ for a typical Julia set.) Therefore sophisticated algorithms are necessary to decide which branches of the tree-like structure in $Or^-(\bar{x})$ should be chosen for an effective picture generation. Such algorithms have been developed and used for our images. Property (2.6) immediately suggests that $J_R$ must be a fractal in many cases. For example, if $R$ has more than two attractive fixed points $a, b, c, \ldots$ then (2.6) implies

$$\partial A(a)=J_R=\partial A(b)=J_R=\partial A(c)=\ldots$$

i.e. the boundaries of all basins of attraction coincide. For example, if $R$ has 3 or 4 attractive fixed points, then $J_R$ is a set of 3-corner-points or 4-corner-points with regard to the respective basins of attraction.

### *The Dynamics Near Indifferent Periodic Points*

Since attractive periodic points belong to $F_R$, and repelling periodic points belong to $J_R$, we may ask about *indifferent points.* This question is in fact very delicate and still not completely understood. Without loss of generality we may assume that $R(0)=0$ and $R'(0)=\lambda$ with $|\lambda|=1$, i.e. $\lambda=\exp(2\pi i\alpha)$ with $\alpha\in[0,1]$. Let

$$R(x)=\lambda x+a_2x^2+a_3x^3+\ldots$$

be the power series for $R$. There are two types of indifferent points: the fixed point 0 is called *rationally* indifferent if $\alpha$ is a rational number; it is called *irrationally indifferent* if $\alpha$ is an irrational number. A rationally indifferent fixed point (or cycle) is also called *parabolic.*
It was known to *Julia* and *Fatou* that

(2.9) $x_0\in J_R$ if $x_0$ is a parabolic periodic point of $R$.

Moreover, they knew that in this case $A(\gamma)\neq\varnothing$, $\gamma=\{x_0, R(x_0),\ldots, R^{n-1}(x_0)\}$ and $\gamma\subset\partial A(\gamma)$. Figures 6, 8, and 9 illustrate this situation.
A more comprehensive characterization of indifferent points requires a closer look into the *dynamics* of $R$ near $R(0)=0$. The following result from [Cam] accounts for the parabolic case:

#### The Parabolic Case

(2.10) Let $\lambda=R'(0)$, $\lambda^n=1$ and $\lambda^k\neq1$ for $0<k<n$. Then either $R^n$ is the identity or there exists a homeomorphism $h$ (defined in a neighborhood of 0) with $h(0)=0$ and

$$h\circ R\circ h^{-1}(x)=\lambda x(1+x^{kn})$$

for some $k\geqslant1$.

#### Siegel Disks

The irrational case is considerably more difficult. We need the concept of stability:

(2.11) $R(0)=0$ is called *stable* if for any neighborhood $U$ of $0$ there exists a neighborhood $V$ of $0$ such that $V\subset U$ and

$$R^k(V)\subset U$$

for any $k\geqslant1$.

Attractive fixed points are obviously stable. To describe the stability of indifferent fixed points we use a result of J. Moser and C.L. Siegel [MS]:

(2.12) Let $R(x)=\lambda x+a_2x^2+\ldots$, $|\lambda|=1$ and $\lambda^n\neq 1$ for any $n\in\mathbb{N}$. Then 0 is a stable fixed point if and only if the functional equation

$$\Phi(\lambda x)=R(\Phi(x))$$

has an analytic solution in a neighborhood of 0.

The above functional equation is called Schröder's equation in honor of E. Schröder who studied its solvability in 1871. What does it actually mean? Assume that it holds for a $\lambda=\exp(2\pi i\alpha)$, then

$$\lambda x=\Phi^{-1}(R(\Phi(x))).$$

I.e. $R$ is *locally equivalent* (or conjugate) to a *rotation* by $2\pi\alpha$. To solve Schröder's equation one tries an Ansatz for $\Phi$: let $R(x)=\lambda x+a_2x^2+\ldots$. If

(2.13) $\Phi(x)=x+b_2x^2+\ldots$

then (2.12) yields

(2.14) $$\sum_{i=2}^{\infty}(\lambda^i-\lambda)b_ix^i=\sum_{i=2}^{\infty}a_i\left(x+\sum_{k=2}^{\infty}b_kx^k\right)^i$$

and formally one can obtain the $b_i$, $i=2,3,\ldots$, by comparing coefficients. The problem then however, is to show convergence. Obviously, this method does not work if $\lambda$ is a root of unity, i.e. $\lambda=\exp(2\pi ip/q)$. When $\alpha$ is irrational, then (2.14) is called a *small divisor problem*. In 1917 G.A. Pfeifer gave an example where Schröder's series (2.13) does not converge. In 1938 H. Cremer was able to provide a whole class of examples for which (2.13) diverges:

(2.15) $\{\lambda: |\lambda|=1 \text{ and } \liminf |\lambda^n-1|^{1/n}=0\}$.

Then in 1942 C.L. Siegel [Si] in a groundbreaking work (which eventually became all important in the Kolmogoroff-Arnold-Moser theory) showed that Schröder's series actually converges provided that $\alpha$ satisfies a diophantine condition ($\lambda=\exp(2\pi i\alpha)$):

(2.16) There exist $\varepsilon>0$ and $\mu>0$ such that

$$\left|\alpha-\frac{m}{n}\right|>\frac{\varepsilon}{n^\mu}$$

for all integers $m$ and positive integers $n$.

The condition on $\alpha$ loosely says *$\alpha$ is badly approximated by rational numbers.* If we write the irrational number $\alpha$ in its continued fraction expansion, we can make this statement precise. Let $\alpha=(a_0, a_1, a_2, \ldots)$ be this expansion, i.e., $a_k\in\mathbb{N}$ and

$$\alpha=a_0+\cfrac{1}{a_1+\cfrac{1}{a_2+\cfrac{1}{a_3+.}}}$$

Then set $\frac{p_n}{q_n}=(a_0, a_1, \ldots, a_n, 0, 0, 0, \ldots)$. These numbers are the best rational approximants to $\alpha$ and it is well known that

$$\frac{1}{(a_{n+1}+2)q_n^2}<\left|\alpha-\frac{p_n}{q_n}\right|<\frac{1}{a_{n+1}q_n^2}.$$

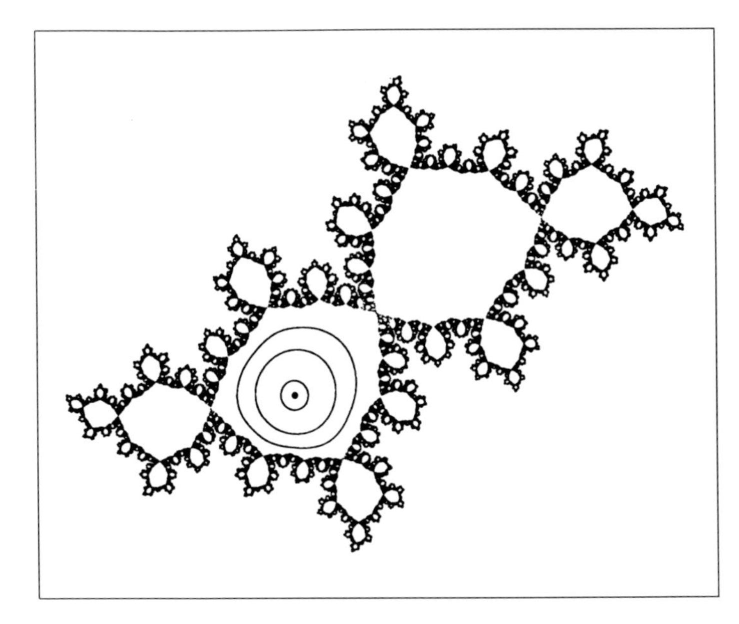

*Fig. 22. Siegel disk around the irrationally indifferent fixed point $x_0 = 0$ for the mapping $x \mapsto x^2 + \lambda x$, $\lambda = e^{2\pi i\alpha}$, $\alpha = (\sqrt{5} - 1)/2$. On the invariant curves (3 are shown) the dynamics is equivalent to a rotation about the angle $\alpha$*

(Here best means that no rational $p/q$ with $q \leqslant q_n$ is closer to $\alpha$ than $\frac{p_n}{q_n}$.)
Hence, for example, if the $a_k$'s stay bounded, one can verify the diophantine condition in (2.16). It is known and not difficult to prove that any algebraic number of degree 2 has such a continued fraction expansion and thus such numbers $\alpha$ (and in fact all algebraic numbers) satisfy the diophantine condition. A prominent example is the golden mean

$$\alpha = (\sqrt{5} - 1)/2, \quad \alpha = (0, 1, 1, 1, \ldots).$$

It is known that the set of $\alpha \in [0, 1]$, for which Siegel's condition is satisfied is a set of full measure. If Schröder's series (2.13) converges, one says that $R$ is linearizable in 0. The maximal domain $D(0)$, containing $R(0) = 0$, in which $\Phi(\lambda x) = R(\Phi(x))$ holds is called a Siegel disk. Maps 22 and 25 show an example from the quadratic family $x \mapsto x^2 + c$. These color maps show a Siegel disk, its preimages under R, and the basin of attraction of $\infty$. The coloring in Map 25 reveals the invariant circles to which the dynamics near the irrationally indifferent fixed point is confined. Figure 22 shows such a fixed point, some invariant curves and the Julia set.
In this example we observe that the critical point $x_c = -e^{2\pi i\alpha}/2$ belongs to the Julia set confirming a recent result of M. Herman (see also Special Section 3). Also note that

(2.17) $\quad x_0 \in F_R$, if $x_0$ is the center of a Siegel disk.

In 1972 H. Rüßmann was able to extend Siegel's result even to certain Liouville numbers. (Liouville numbers $\lambda$ are very close to rational numbers.) The

exact condition on $\alpha$ for which Schröder's series converges is an open and apparently very deep problem. Consequently, one cannot as of yet always determine if a given irrationally indifferent fixed point is in either the Julia set or the Fatou set.
If $R(x)=x^2+\lambda x$, then of course for every $\lambda \in S^1$ we have that $x_0=0$ is an indifferent fixed point. Imagine that $\lambda$ is varying on $S^1$. Then, no matter how small the change is, the dynamics near $x_0=0$ will undergo most dramatic changes. This is because any change of $\lambda$ will always result in infinitely many parabolic and Siegel disk cases.

### *The Hausdorff Dimension*

According to B. B. Mandelbrot a set $X$ is called a *fractal* provided its *Hausdorff dimension* $h(X)$ is not an integer. Intuitively $h(X)$ measures the growth of the number of sets of diameter $\varepsilon$ needed to cover $X$, when $\varepsilon \to 0$. More precisely, if $X \subset \mathbb{R}^m$, let $n(\varepsilon)$ be the number of $m$-dimensional balls of diameter $\varepsilon$ needed to cover $X$. Then if $n(\varepsilon)$ increases like

(2.18) $\quad n(\varepsilon) \propto \varepsilon^{-D}$ as $\varepsilon \to 0$,

one says that $X$ has Hausdorff dimension $D$. It is not hard to show that if $C$ is the familiar Cantor set, then

$$h(C) = \log 2/\log 3.$$

A rigorous definition for $h(X)$ proceeds as follows: let $X$ be a subset of a metric space and let $d>0$. The $d$-dimensional *outer measure* $m_d(X)$ is obtained from

$$(2.19)\quad \begin{cases} m_d(X,\varepsilon) = \inf\{\sum_{i\in I} (\operatorname{diam} S_i)^d\}, \text{ where the inf is over all finite coverings} \\ \text{of } X \text{ by sets } S_i \text{ with diameter less than } \varepsilon>0. \\ m_d(X) = \lim_{\varepsilon\to 0} m_d(X,\varepsilon). \end{cases}$$

Now $m_d(X)$, depending on the choice of $d$, may be finite or infinite. F. Hausdorff showed in 1919 that there is a unique $d=d^*$ at which $m_d(X)$ changes from infinite to finite as $d$ increases. This then leads to the definition

(2.20) $\quad h(X) = \sup\{d \in \mathbb{R}_+ : m_d(X) = \infty\}$.

(See [Fal] for more details and examples.)
Recently, D. Ruelle [Ru2] obtained the following remarkable result: let $J_c$ be the Julia set for $x \mapsto x^2+c$. Then for $|c| \ll 1$ one has

$$(2.21)\quad h(J_c) = 1 + \frac{|c|^2}{4\log 2} + \textit{higher order terms}.$$

It is also known (see [Bro]) that for $c$ small $J_c$ is a Jordan curve (i.e. the homeomorphic image of the unit circle). In fact $J_c$ is a Jordan curve for any $c$ in the main part (the cardioid) of the Mandelbrot set. Though Julia sets are typically of fractal nature, almost nothing is known about their Hausdorff dimension. Ruelle's result seems to be the first sharp result in that direction.

*Julia Sets for Transcendental Maps*

In Section 6 we will continue to discuss some special classes of Julia sets, namely those for Newton's method in the complex plane. In Section 7 we compare our results from Section 6 with some first findings for Julia-like sets obtained from Newton's method for real equations. Our experiments there reveal structures which look quite different from the baroque structures which we have seen so far for rational mappings in the complex plane. One is tempted to attribute this apparent baroqueness to the underlying complex analytic structure. This is a little premature, however, as the following example indicates. R. Devaney [De] recently has studied some first examples of transcendental mappings, as for example

(2.22) $E_\lambda(x) = \lambda \exp(x)$,
(2.23) $S_\lambda(x) = \lambda \sin(x)$,

$\lambda \in \mathbb{C}$, in $\overline{\mathbb{C}}$. He obtained some very remarkable results, a few of which we want to include here. Defining the Julia set for $E_\lambda$ (or $S_\lambda$) according to (2.1), i.e.

(2.24) $J_\lambda = \text{closure}\,\{x \in \mathbb{C}: x \text{ is a periodic repelling point of } E_\lambda \text{ (resp. } S_\lambda)\}$

then a first result is

(2.25) $J_\lambda = \text{closure}\,\{x \in \mathbb{C}: E_\lambda^n(x) \to \infty \text{ (resp. } S_\lambda^n(x) \to \infty) \text{ as } n \to \infty\}$.

Note that there is a distinctive change in the behavior of $E_\lambda$, for example, as $\lambda$ passes through $1/e$ along the real axis (see Fig. 23).
If $\lambda < 1/e$ there is an attractive fixed point $Q_\lambda$ and a repelling fixed point $P_\lambda$, while if $\lambda > 1/e$ there is none. Also, if $\lambda < 1/e$ one has apparently that

(2.26) $\{x \in \mathbb{R}: x \geq P_\lambda\} \subset J_\lambda$.

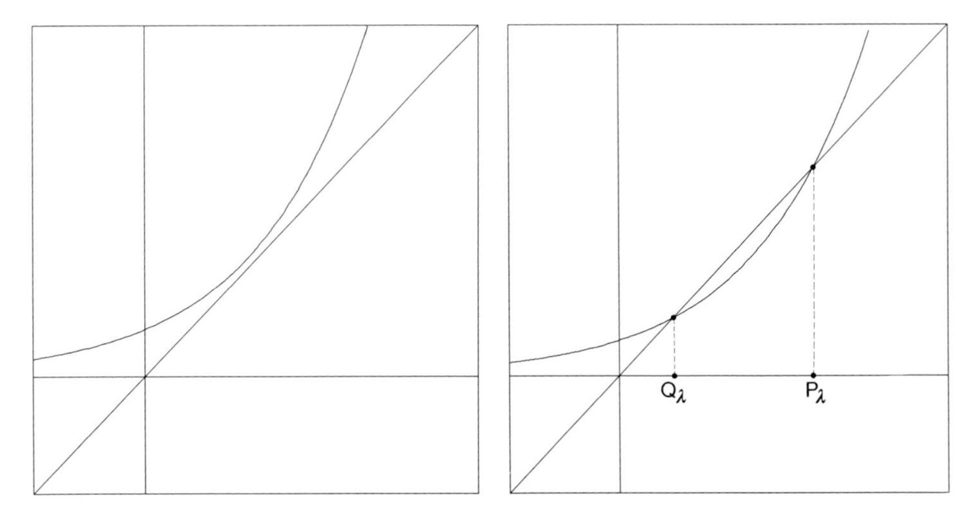


*Fig. 23. The graph of* $x \mapsto \lambda \exp(x)$, $x \in \mathbb{R}$, $\lambda \in \mathbb{R}$. Left: $\lambda > 1/e$; right: $\lambda < 1/e$

Devaney calls this ray a hair. It turns out to be crucial in the description of $J_\lambda$. One of his main results is:

$$(2.27)\begin{cases}\text{If } \lambda > 1/e, \text{ then } J_\lambda = \mathbb{C}.\\ \text{If } \lambda < 1/e, \text{ then } J_\lambda \text{ is a nowhere dense Cantor set of curves which}\\ \text{form the boundary of a single basin of attraction.}\end{cases}$$

Thus, as $\lambda$ increases and passes through $1/e$, $J_\lambda$ experiences an explosion. Pictures of Julia sets for this family are quite difficult to obtain and we refer to [De] for a sketch of a picture.
Inspired by Mandelbrot's work (see Special Section 4) Devaney also discusses a bifurcation set for $E_\lambda$ in the $\lambda$-plane:

(2.28) $B = \{\lambda \in \mathbb{C} : J_\lambda = \mathbb{C}\}$ and
(2.29) $C = \mathbb{C} \setminus B$.

He shows that the interior of C contains components which are indexed by the period of attractive periodic points. Figure 24 gives a sketch of B in black.
This study is based on another result of Devaney:

(2.30) If $E_\lambda^n(0) \to \infty$, then $J_\lambda = \mathbb{C}$.

This result suggests to color a point in the $\lambda$-plane black if $|E_\lambda^n(0)| \geq M$, $M \gg 1$, for some $n < N_{\max}$. The picture in Fig. 24 has to be interpreted with some care. For example, the white set contains points like $\lambda = 2k\pi i$. I.e., $E_\lambda^2(0) = E_\lambda(\lambda) = \lambda$. In general:

(2.31) If 0 is preperiodic (i.e. $E_\lambda^n(0)$ is periodic for some $n \geq 1$), then $J_\lambda = \mathbb{C}$.

*Fig. 24. Mandelbrot-like set for $E_\lambda$ (the solid black domains are an artifact of the low iterational resolution $N_{max} = 90$)*

For each $\lambda$ such that $|E_\lambda^n(0)| \geq M$ for some $n \leq N_{\max}$, there is a first $n$ of this kind. This associates an index to each such $\lambda$, which can be used to determine a color in a *Color Look Up Table* to generate beautiful color graphics.
Comparing Devaney's results with pictures of other Julia sets, they seem to share more properties with the Julia-like sets in Section 7 than with Julia sets of rational mappings. For example, $J_\lambda$ is not locally connected, like many of the Julia-like sets in Section 7. Also we have found that the Julia-like sets there typically contain "hairs" like (2.26). ($X \subset \mathbb{C}$ is called locally connected if for $U$ open in $\mathbb{C}$ and $U \cap X \neq \emptyset$ one has that for any $x \in U \cap X$ there is a neighborhood $V \subset U$, $x \in V$, such that $V \cap X$ is connected.)

*Generating Pictures of Julia Sets*

There are essentially two different ways to generate pictures of Julia sets. One is based on (2.5) and the other builds on (2.6). None of the methods has an advantage over the other. In some cases the first method is very successful while the second is quite unsatisfactory, and vice versa. Then there is a large class of cases in which both work fine. But there is also a large class of Julia sets for which it is very difficult if not impossible to generate satisfactory pictures. This class contains Julia sets bounding parabolic domains, i. e. the mapping has a parabolic periodic point.

The Inverse Iteration Method (IIM)

Given a rational mapping $R$ and a periodic repeller $\bar{x} \in J_R$, property (2.5) suggests to compute

(2.32) $J_R^n = \{x \in \mathbb{C} : R^k(x) = \bar{x} \text{ for some } k \leq n\}$.

Since $J_R = \text{closure} \left( \bigcup_{n \geq 0} J_R^n \right)$ one expects that plotting $J_R^n$ for a sufficiently large $n$ should give a good picture of $J_R$. Indeed, if $J_R^n$ is uniformly distributed over $J_R$, then this method generates a satisfactory picture of $J_R$. Intuitively speaking, we say that the inverse orbit $Or^-(\bar{x})$ distributes uniformly provided the number of points in ($\varepsilon > 0$, $\varepsilon$ small)

$$J_R^n \cap D(x,\varepsilon), \qquad (x \in J_R,\ D(x,\varepsilon) = \{y : |y-x| < \varepsilon\})$$

is essentially independent of $x$ for large $n$. Unfortunately, this is not typical. More typically one has neighborhoods on $J_R$ which are visited only extremely rarely. In these cases the direct IIM is inappropriate. Recall that the number of elements in $J_R^n$ grows like $d^n$, where $d$ is the degree of $R$. Figure 25 shows a typical situation.
The shortcomings of this experiment become apparent if one compares Fig. 25 with Map 18, where the red domain identifies the basin of attraction of an attractive cycle of period 11 and the Julia set is its boundary. Figures 26 and 27 show another Julia set from the quadratic family $R(x) = x^2 + c$ together with a demography of $J_R^n$ based on a covering of $J_R$.
In these experiments, $J_R$ is put on a square lattice in $\mathbb{C}$ with small mesh size. The number of points from $J_R^n$ in each little box of that lattice is measured and

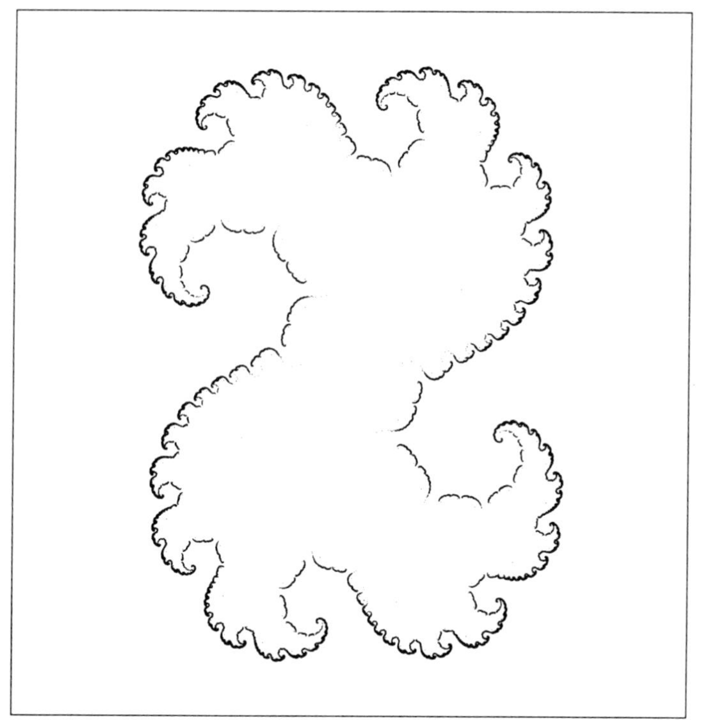

*Fig. 25.* $J_R^n$, $n=22$; $R(x)=x^2+c$ and $c=0.32+0.043\,i$; obtained by IIM

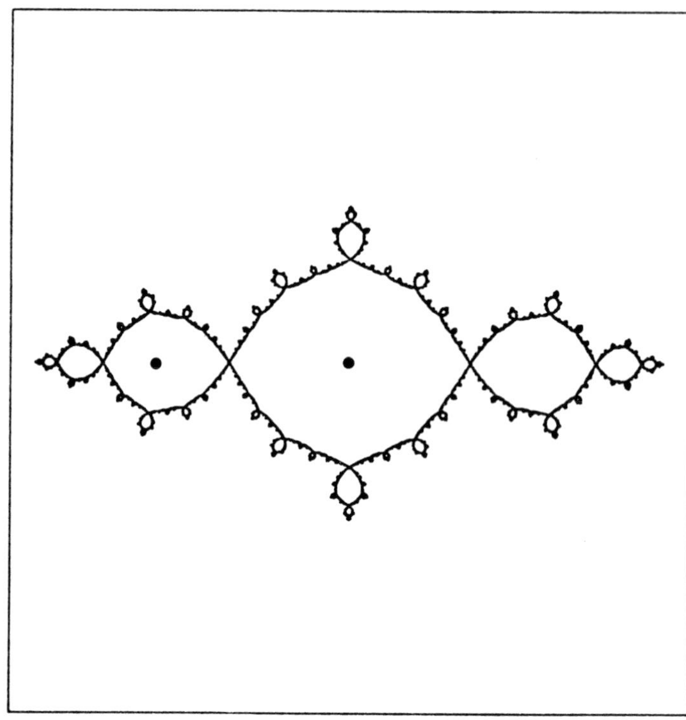

*Fig. 26.* $J_R$, $R(x)=x^2+c$ and $c=-1$; $\{0,-1\}$ is a superattractive cycle

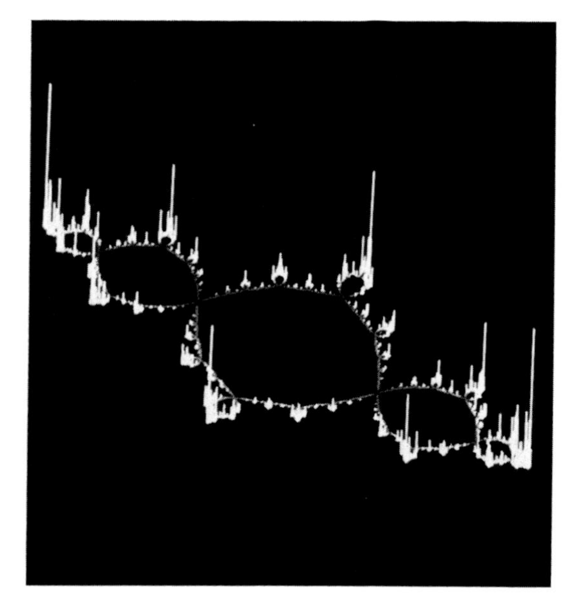

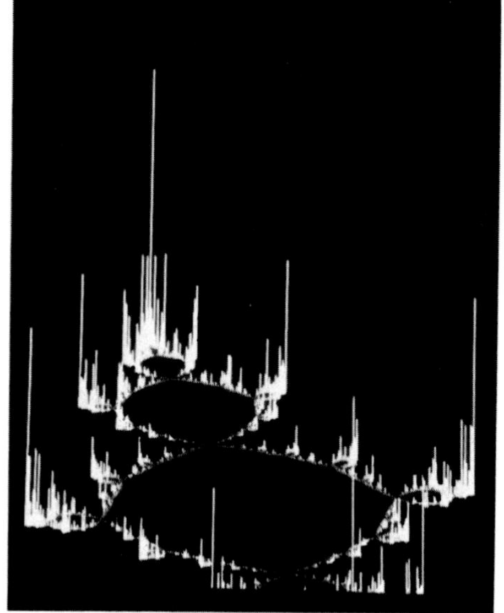

*Fig. 27.* $J_R^n$, $n=21$; $R(x)=x^2+c$ and $c=-1$; distribution of typical inverse orbit $Or^-(\bar{x})$, $\bar{x}=1/2(1+\sqrt{5})$, displayed in vertical bars

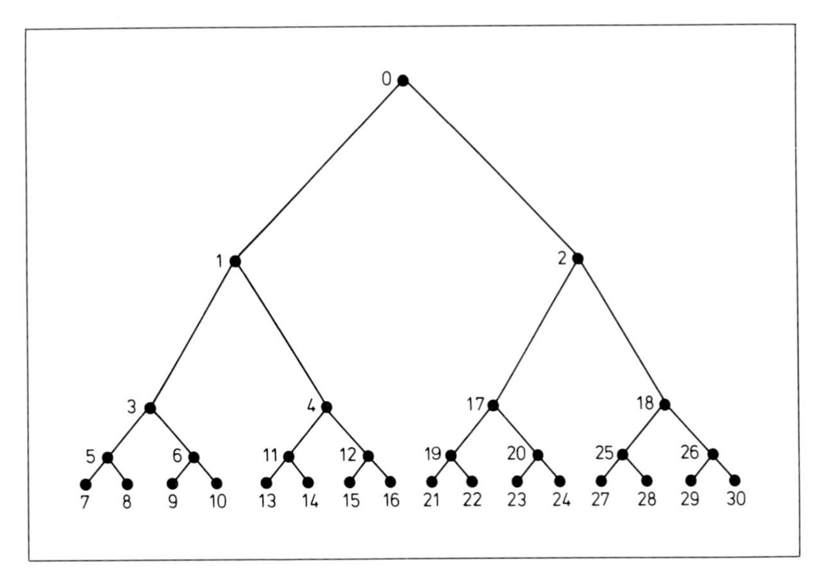


*Fig. 28. Storage efficient hierarchy in the inverse orbit of a mapping of degree* 2

represented by a vertical bar, thus visualizing the distribution. Apparently, the tips of $J_R$ are visited most frequently, while the branch points seem to be avoided most often. Nevertheless, for $n=21$, $J_R^n$ creates a dense enough set of points, so that the nonuniformity has little effect on Fig. 26. But it is this effect which leaves Fig. 25 so unsatisfactory.

Besides these shortcomings the IIM cannot be applied without some appropriate data base management, because of the growth rate $d^n$. For example, Fig. 28 shows the tree structure in $J_R^n$ *for* $d=2$.

Obviously, there are other ways to index the different levels $J_R^k$, $k=0, 1, \ldots, n$. Straightforward indexing by rows means that for the computation of level $(n+1)$ one needs all $d^n$ elements of $J_R^n$, which easily exceeds acceptable storage requirements. If, however, one anticipates that $n$ iterations suffice, then there is an obvious way to index the tree, which only requires $d\cdot(n-1)$ units of storage for the generation of $J_R^n$ (see Fig. 28).

In general the distribution of $Or^-(\bar{x})$ is characterized by a measure $\mu_R$ having support $J_R$: According to M.J. Ljubich [Lj] one has

(2.33) There is an invariant Borel measure $\mu_R$ with support $J_R$. For any $x \in \overline{\mathbb{C}} \setminus E$ ($E$ is some exceptional set) $\mu_R$ is the weak limit of the $\mu_{m,x}$ where

$$\mu_{m,x} = \frac{1}{d^m} \sum_{z \in R^{-m}(x)} \delta_z \quad \text{and} \quad \delta_z(u) = \begin{cases} 1, & u=z \\ 0, & \text{else.} \end{cases}$$

If one knew $\mu_R$ in advance, then it would be obvious how to modify IIM. The following strategy takes account of $\mu_R$ in some approximative sense, without actually knowing it.

The Modified Inverse Iteration Method (MIIM)

Intuitively, the idea is very simple. Points in $J_R^k$, $k<n$, which belong to parts carrying a lot of measure with respect to $\mu_R$ should be used much less frequently in the inverse iteration than they actually occur. In fact the number of them going into the iteration process should be inversely proportional to $\mu_R$. The strategy is: put $J_R$ on a square lattice with small mesh size $\beta$. Then, for any box $B$ of that mesh, stop using points from $B$ for the inverse iteration, provided a certain number $N_{max}$ of such points in $B$ have been used. Optimal choices of $\beta$ and $N_{max}$ naturally will depend on $\mu_R$ and on computergraphical parameters, such as the resolution of the given system. Therefore an interactive and adaptive algorithm is desirable. Our student H.-W. Ramke has developed such algorithms which in many cases turned out to be highly efficient and satisfactory. In fact, most of our black/white pictures of Julia sets in this book were generated with his algorithms.
In some cases, however, where MIIM fails one can obtain satisfactory results by using algorithms of much less sophistication, which are based on (2.6).

The Boundary Scanning Method (BSM)

Recall that if $R$ has an attractive fixed point $a$, i.e., $|R'(a)|<1$, $R(a)=a$, and $A(a)$ is its basin of attraction, then

$$J_R=\partial A(a).$$

Now, if $R$ has more than one attractive fixed point, say $a$ and $b$, there is a rather simple way to generate a picture of $\partial A(a)=J_R=\partial A(b)$:
One chooses a square lattice covering some region of $\overline{\mathbb{C}}$ in which $J_R$ has to be generated. Let $B$ be a typical open box in that mesh of width $\beta$. Assume that $B\cap J_R\neq\emptyset$. Then $B$ must contain points both from $A(a)$ and $A(b)$.
This suggests the following algorithm: let $N_{max}$ and $0<\varepsilon\ll 1$ be given and let $B$ be any box in the lattice with corner points $c_1$, $c_2$, $c_3$ and $c_4$. One tests

- if all corner points belong to the same basin,
  or
- if the corner points belong to different basins.

In the last case one colors $B$ black, for example, and in the first case white. Numerically it then remains to decide whether $c_i$ belongs to $A(a)$, for example, or not.

$$-\begin{cases}\text{If for some } k\leq N_{max} \text{ it is found that } |R^k(c_i)-a|<\varepsilon, \text{ then } c_i\in A(a).\\ \text{If not one decides } c_i \text{ is not in } A(a).\end{cases}$$

(Obviously, the last decision introduces an error of resolution.) Again the parameters $\beta$, $\varepsilon$ and $N_{max}$ depend strongly on $R$ and it may turn out that there is no choice which leads to a satisfactory picture of $J_R$. The requirement of the existence of at least two attractive fixed points can be relaxed, given some further information about $R$. For example, one could test a basin of attraction (of an attractive or parabolic point) versus a Siegel disk or versus a Herman ring or even versus a set on which $R$ diverges (e.g. $x\mapsto x^2+c$, $c=i$, $\overline{\mathbb{C}}=A(\infty)\cup J$

and $J$ is a dendrite). Obviously, if $R$ has an attractive periodic point $a$ of period $m$, then $R^k(a)$, $k=0, 1, \ldots, m-1$ are attractive fixed points for $R^m$.

## Principal Numerical Difficulties with Parabolic Fixed Points

Experimental experience shows that generating Julia sets in the presence of parabolic points is a particularly hard problem. In [Bl] (Fig. 3.16, p. 102), for example, one finds an experiment which is supposed to show the Julia set of

$$R_0(x)=\lambda x+x^2, \; \lambda=\exp(2\pi i/20).$$

Note that $x_0=0$ is a parabolic fixed point. The picture in [Bl], however, is quite misleading. It was presumably generated by an IIM-algorithm. Figure 29 shows our result obtained by an appropriately tuned BSM-algorithm. The difference to [Bl] is quite dramatic, but still Fig. 29 has several shortcomings. Firstly, we should see $x_0=0\in J_{R_0}$ (according to (2.9)). Secondly, ($\lambda^{20}=1$) the Julia set should approach $x_0=0$ in 20 different directions (between 20 petals). In fact one can see hints of these directions, but apparently the Julia set seems to stop at a certain distance from $x_0$. This effect is numerical in nature, i.e. quite different from the effects due to intrinsic nonuniformities in the invariant measure.
Let *eps* be the machine unit round off error of a given computer (i.e. *eps* is the largest machine number so that $1+eps=1$, in machine arithmetic). Furthermore let $R$ satisfy the assumptions of (2.10). In particular,

$$h\circ R\circ h^{-1}(x)=\lambda x(1+x^{kn}).$$

Then if $|x|<eps^{1/kn}$ one obviously (in machine arithmetic) cannot distinguish $\lambda x$ from $\lambda x(1+x^{kn})$, and hence $R^n(h^{-1}(x))$ cannot be distinguished from $h^{-1}(x)$. The following table illustrates this unfortunate effect by giving the distance $r$ from 0, at which iteration simply stops.

| $kn$ | $r$ | |
|---|---|---|
| | $eps=10^{-8}$ | $eps=10^{-16}$ |
| 2 | $1\cdot10^{-4}$ | $1\cdot10^{-8}$ |
| 5 | $2.51\cdot10^{-2}$ | $6.309\cdot10^{-4}$ |
| 20 | $3.981\cdot10^{-1}$ | $1.584\cdot10^{-1}$ |
| 100 | $8.318\cdot10^{-1}$ | $6.918\cdot10^{-1}$ |

This explains the shortcoming of Fig. 29 ($kn=20$), but also, why we do not see much of this effect in Fig. 26 ($kn=2$). Figure 30 shows how Fig. 29 should have looked. It was obtained by an MIIM-algorithm applied to

$$R_\varepsilon(x)=(1+\varepsilon)\lambda x+x^2, \; \lambda=\exp(2\pi i/20).$$

For $\varepsilon$ changing from 0 to a small positive number ($\varepsilon=0.001$ in Fig. 30) the parabolic fixed point $x_0=0$ bifurcates into an attractive cycle of period 20, and $x_0$ becomes a repelling fixed point.

Fig. 29. *Julia set for* $x \mapsto \lambda x + x^2$, $\lambda = e^{2\pi i/20}$; $x_0 = 0$ *is a parabolic fixed point*

Fig. 30. *Julia set for* $x \mapsto (1+\varepsilon)\lambda x + x^2$, $\lambda = e^{2\pi i/20}$, $\varepsilon = 0.001$; $x_0 = 0$ *is a repelling fixed point*

*Level Sets and Binary Decompositions*

Our BSM-algorithm suggests a variant which leads to a dynamic decomposition of basins of attraction. Let $R$ be a rational function with an attractive fixed point $a = R(a)$. We define *level sets of equal attraction in* $A(a)$ *with respect to a target set* $L_0(a)$:
Let $a \in L_0(a) \subset A(a)$. Then set

$$L_k(a) = \{x: R^k(x) \in L_0(a) \text{ and } R^l(x) \notin L_0(a) \text{ for } l < k\},\ k = 1, 2, \ldots$$

Typically, one chooses $L_0(a) = \{x: |x-a| \leq \varepsilon\}$ or if $a = \infty$, $L_0(\infty) = \{x: |x| \geq 1/\varepsilon\}$, for some $0 < \varepsilon \ll 1$. In this way each $x \in L_k(a)$ receives an index which can be used to select a certain color from a *Color Look Up Table* when thought of as a *pixel* in a color raster device. All color experiments showing Julia sets, as for example Maps 3–10 or 61–63, are based on such decompositions using 256 colors out of $(256)^3$ choices. In other words, the grading of colors illustrates the dynamic distance from the respective center, the Julia set having infinite distance because $\partial L_k \to J_R$ as $k \to \infty$.
The idea of examining the dynamics by means of level sets is used throughout this book. It was also fundamental in visualizing the potential of the Mandelbrot set in Maps 26–54. There the target set is

$$L_0(\infty) = \{c: |c| > \varepsilon^{-1}\}$$

for some $0 < \varepsilon \ll 1$; except for Maps 53, where

$$L_0(\infty) = \{c: (|\text{Re } c|^p + |\text{Im } c|^p)^{1/p} > \varepsilon^{-1}\}$$

for some $0<p<1$. The equipotential lines of the potential of the Mandelbrot set are approximated by the boundaries of the level sets

$$L_k(\infty)=\{c\in\mathbb{C}: p_c^k(c)\in L_0(\infty) \text{ and } p_c^l(c)\notin L_0(\infty),\ l<k\},\ k=1,2,\ldots$$

where $p_c(x)=x^2+c$ (see Section 4 for details).
The target set $L_0$ can be chosen arbitrarily. In particular, one can further decompose a given $L_0$ into $m$ disjoint subsets, which induce accordingly a decomposition of each level $L_k$. We call these decompositions $m$-adic decompositions. The case of $m=2$ is of particular importance for us. It will help us to discuss the *field lines* of the potential of the Mandelbrot set. These decompositions were introduced in [PSH] to study basins of attraction for Newton's method in $\mathbb{C}$.
We need an old result due to Boettcher, from 1905, see also [Bl]:

(2.34) Let $R$ be a rational mapping and $R(a)=a$. Suppose that $R^{(k)}(a)=0$, $k=1,\ldots,n-1$ and $R^{(n)}(a)\neq 0$ ($R^{(j)}(x)$ is the $j$-th derivative at $x$). Then there are open neighborhoods $U$ of 0 and $V$ of $a$, and there is a conformal mapping $\Phi: U\to V$ such that

$$R(\Phi(u))=\Phi(u^n) \Leftrightarrow \Phi^{-1}\circ R\circ\Phi(u)=u^n.$$

In other words $R$ is locally equivalent to $u\mapsto u^n$.

## Binary Decomposition for $x\mapsto x^2$

Now let $x=\exp(2\pi i\alpha)$, $0\leq\alpha\leq 1$, and let $r(\alpha)$ be the ray through 0 and $x$ in $\mathbb{C}$. Then $r(\alpha)$ decomposes into two parts: the part $r_0(\alpha)$ inside the unit circle, and $r_\infty(\alpha)$ outside the unit circle. We call $r_0(\alpha)$ the *internal angle* of $x$ and $r_\infty(\alpha)$ the *external angle* of $x$. The dynamics of $R(x)=x^2$ in $\mathbb{C}$ provides a nice way to visualize angles of the form

(2.35) $(\alpha_0+k/2^n) \bmod 1$,

$\alpha_0$ fixed, $k, n\in\mathbb{N}$. Recall that if $\alpha=p/q$, $p$ and $q$ relatively prime and $0<\alpha<1$, then the binary expansion

$$\alpha=\sum_{k=1}^{\infty} a_k 2^{-k},\ a_k\in\{0,1\}$$

is obtained in the following way:

(2.36)
- if $q=2^n$ for some $n$, then the expansion is obvious.
- if $q\neq 2^n$ for all $n$, then

$$a_k=\begin{cases} 0, & \text{provided } 0<(2^{k-1}\alpha) \bmod 1<1/2 \\ 1, & \text{else}\end{cases}$$

Now let $L_0(0)=\{x: |x|\leq\varepsilon\}$ and $L_0(\infty)=\{x: |x|\geq\varepsilon^{-1}\}$ for some $0<\varepsilon\ll 1$. Then $R(x)=x^2$ defines level sets $L_k$ in $A(0)$ and $A(\infty)$. For each $x\in L_k$ one determines a binary coding ($\alpha_o$ is fixed):

$$\text{(2.37)}\quad x \text{ is coded } \begin{cases} 0, & \text{if } (2^k\alpha_0) \bmod 1\leq \frac{1}{2\pi}\arg R^k(x)\leq \left(2^k\alpha_0+\frac{1}{2}\right) \bmod 1 \\ 1, & \text{else.}\end{cases}$$

*Fig. 31. Binary decomposition of $A(0)$ and $A(\infty)$ for $x \mapsto x^2$ (white $\leftrightarrow$ 0, black $\leftrightarrow$ 1)*

Figure 31 shows a *binary decomposition* of $A(0)$ and $A(\infty)$ for $\alpha_0 = 0$ (0 is color coded white and 1 is color coded black).
Our coding (2.37) leads to a cell structure in the $L_k$'s, and naturally the angles (2.35) are identified by those rays which eventually (i.e. close to the unit circle) separate black and white cells. In that way (2.37) pronounces the special angles (2.35). The cell structure in Fig. 31 has also another obvious and useful interpretation. Let $\alpha$ be such that

$$(\alpha - k/2^n) \bmod 1 \neq 0$$

for any $k, n \in \mathbb{N}$. Then $r_0(\alpha)$ is a ray which intersects a unique cell in each $L_k(0)$. Let this cell be $C_k$. One can therefore literally *read off* the binary expansion of $\alpha$:

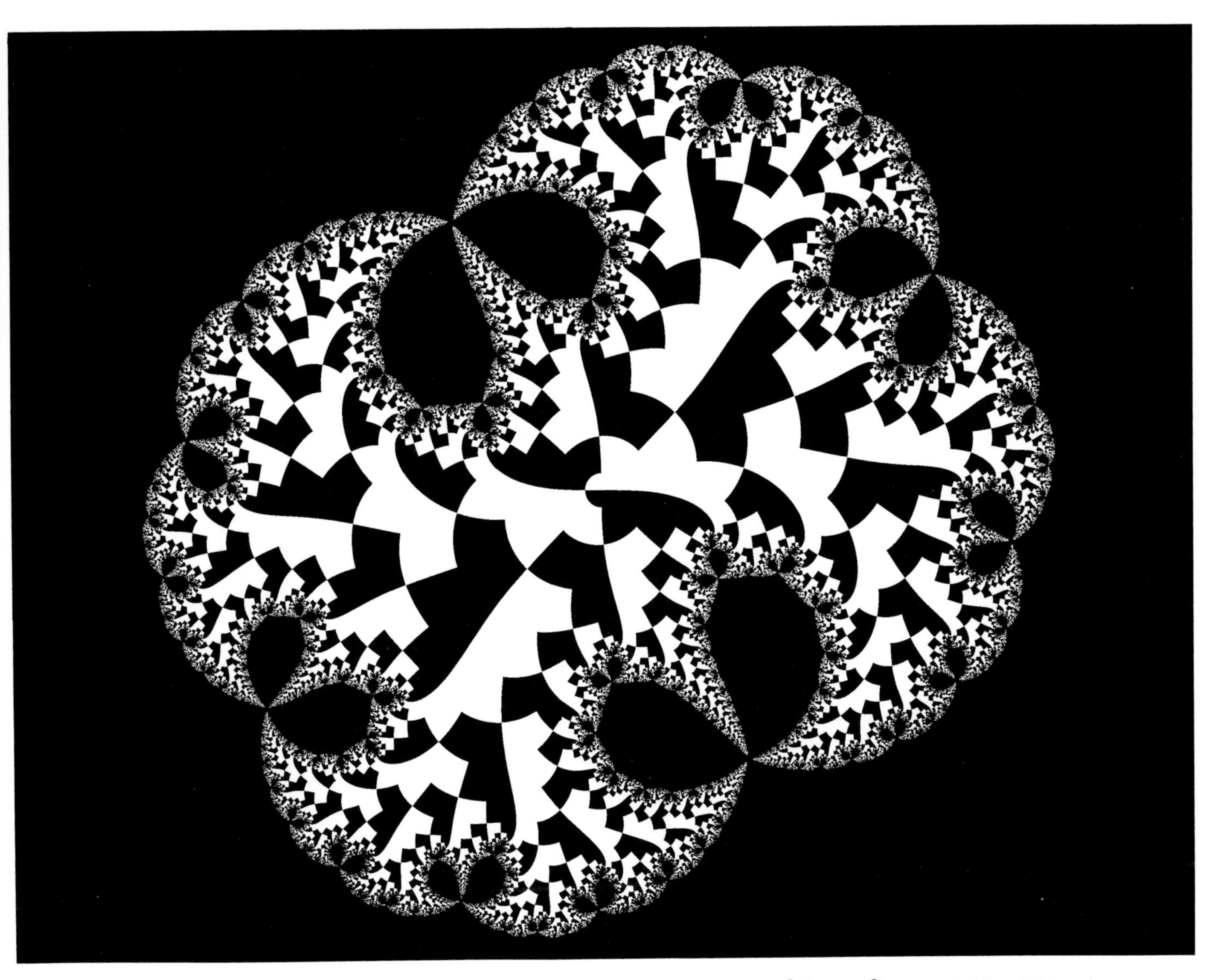

*Fig. 32. Binary decomposition of* $A(0)$ *for* $u \mapsto u^2/(1+cu^2)$, $c=-0.12+0.74\,i$; *this Julia set corresponds to Fig. 4, the "lapin"*

$$(2.38)\quad \begin{cases} a_k = \begin{cases} 0, \text{if } C_{k-1} \text{ is white}, k=1,2,\ldots \\ 1, \text{else} \end{cases} \\ \alpha = \sum_{k=1}^{\infty} a_k 2^{-k} \end{cases}$$

As an application, we consider the mapping $p_c(x)=x^2+c$ once again. The point $x=\infty$ is a superattractive fixed point for all $c$ and we like to look into a binary decomposition of $A(\infty)$. Modulo the coordinate change $x=1/u$ the polynomial $p_c$ is equivalent to

$$R_c(u)=\frac{u^2}{1+cu^2},$$

which has the superattractive fixed point $u=0$ for all $c$. Hence, as a consequence of (2.34), in a neighborhood of 0 $R_c$ is equivalent to $z \mapsto z^2$. In fact, if $c$ is in the Mandelbrot set, then $R_c$ has no other critical point than $0=1/\infty$ itself in $A(0)$ and therefore the conjugation can be extended to all of $A(0)$. In other words, if one replaces $R$ in (2.37) by $R_c$ for such a $c$, one will obtain a binary decomposition of $A(0)$; which permits us to read off binary addresses for points from $\partial A(0)$ (see Fig. 32). This interpretation follows from (2.34): The conjugation $\Phi$ can be chosen so that $\Phi(0)=0$ and $\Phi'(0)=1$. Then the binary decomposition of $A(0)$ decodes the transformed angles $\Phi(r_0(\alpha))$, where $r_0(\alpha)$ is any internal angle in the unit disk. In other words, $\Phi$ induces a structure of internal angles in $A(0)$. In this sense binary decompositions are useful experimental tools for the identification of external/internal angles (see Special Sections 4 and 5). Further applications are discussed in Section 7.

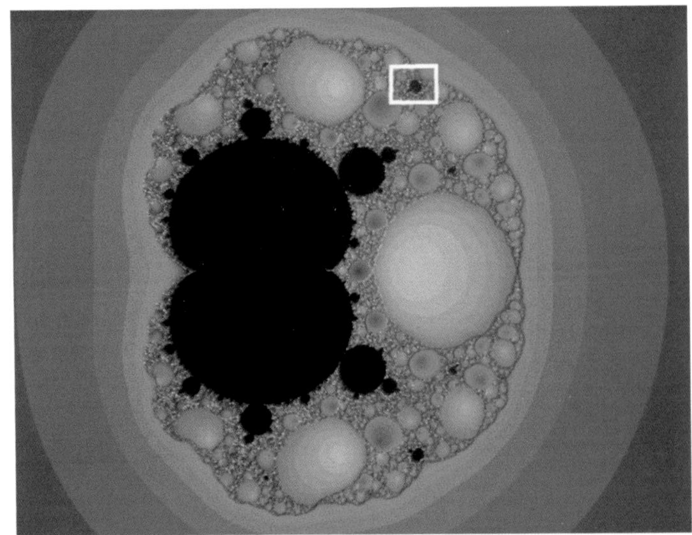

Map 1

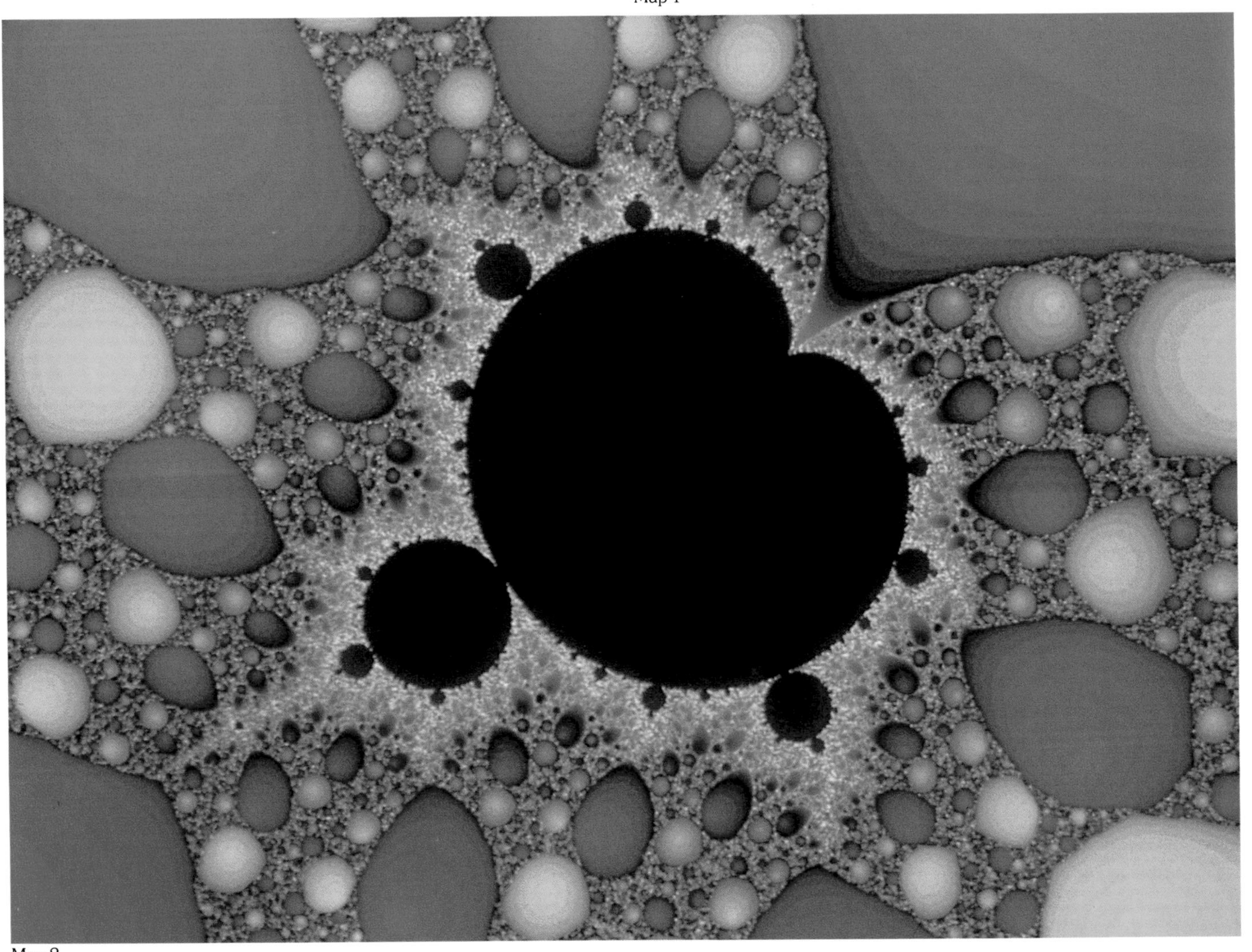

Map 2

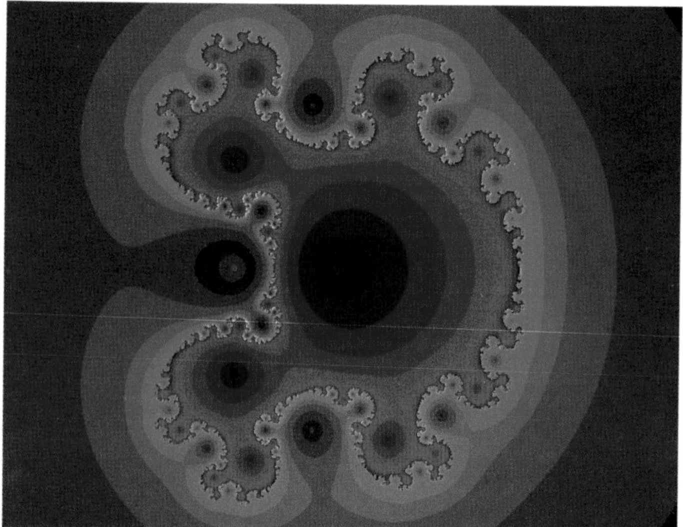
Map 3

Map 4

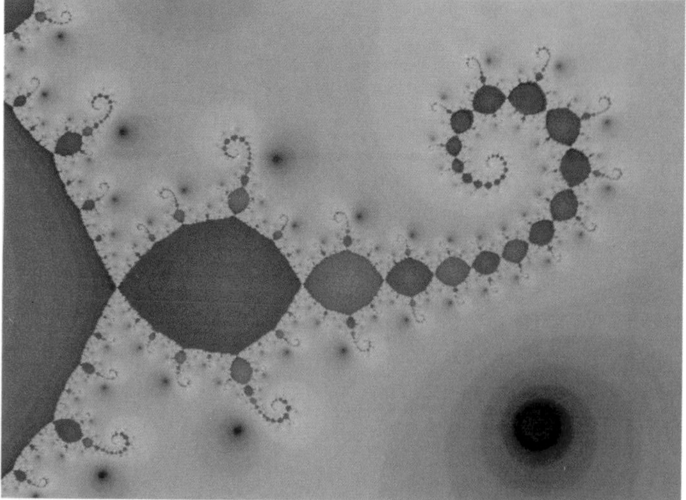
Map 6

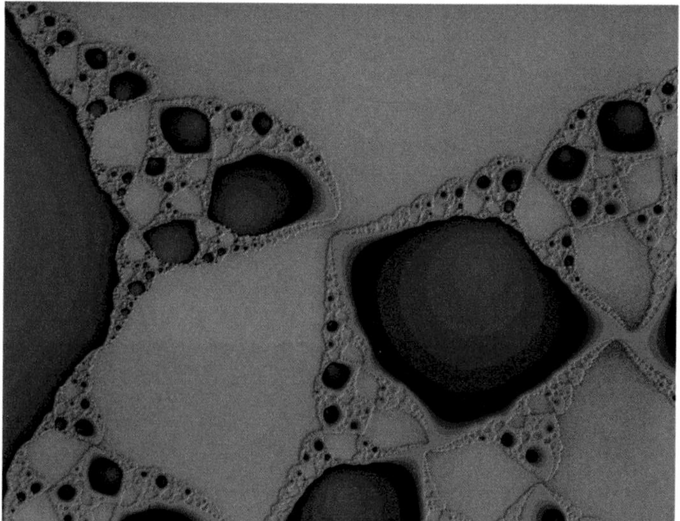
Map 5

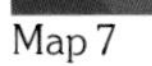

Map 7

Map 8

Map 9

Map 10

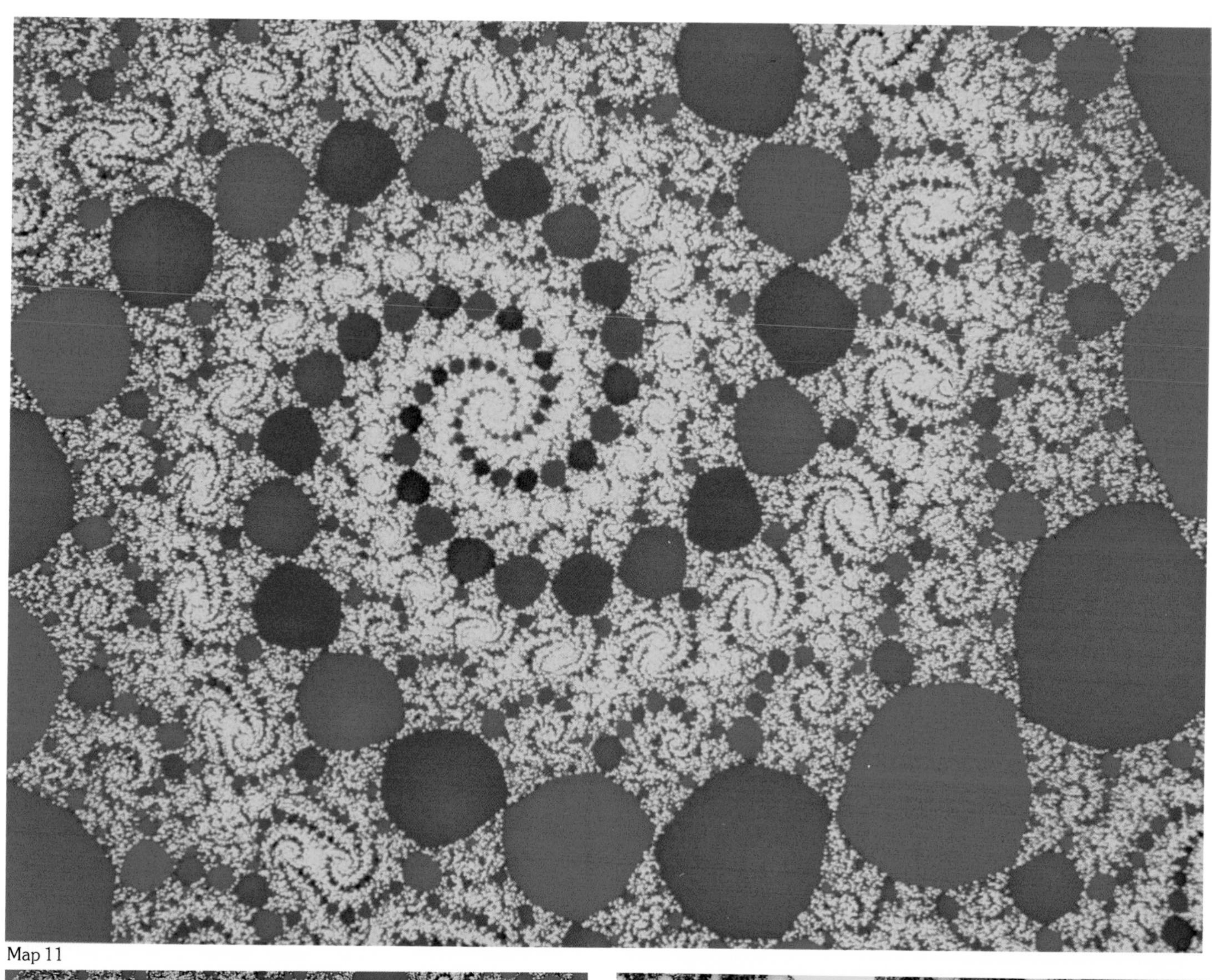

Map 11

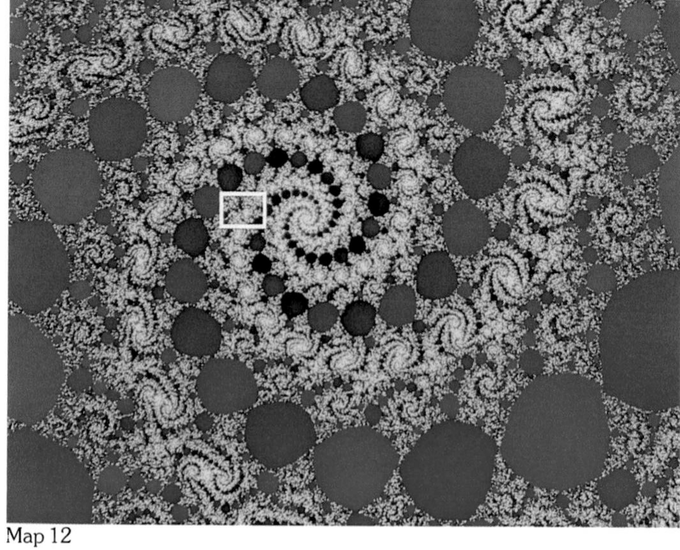

Map 12

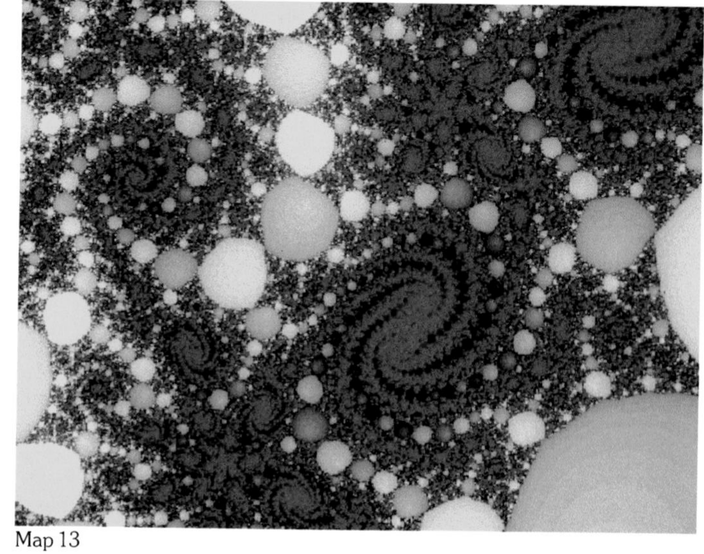

Map 13

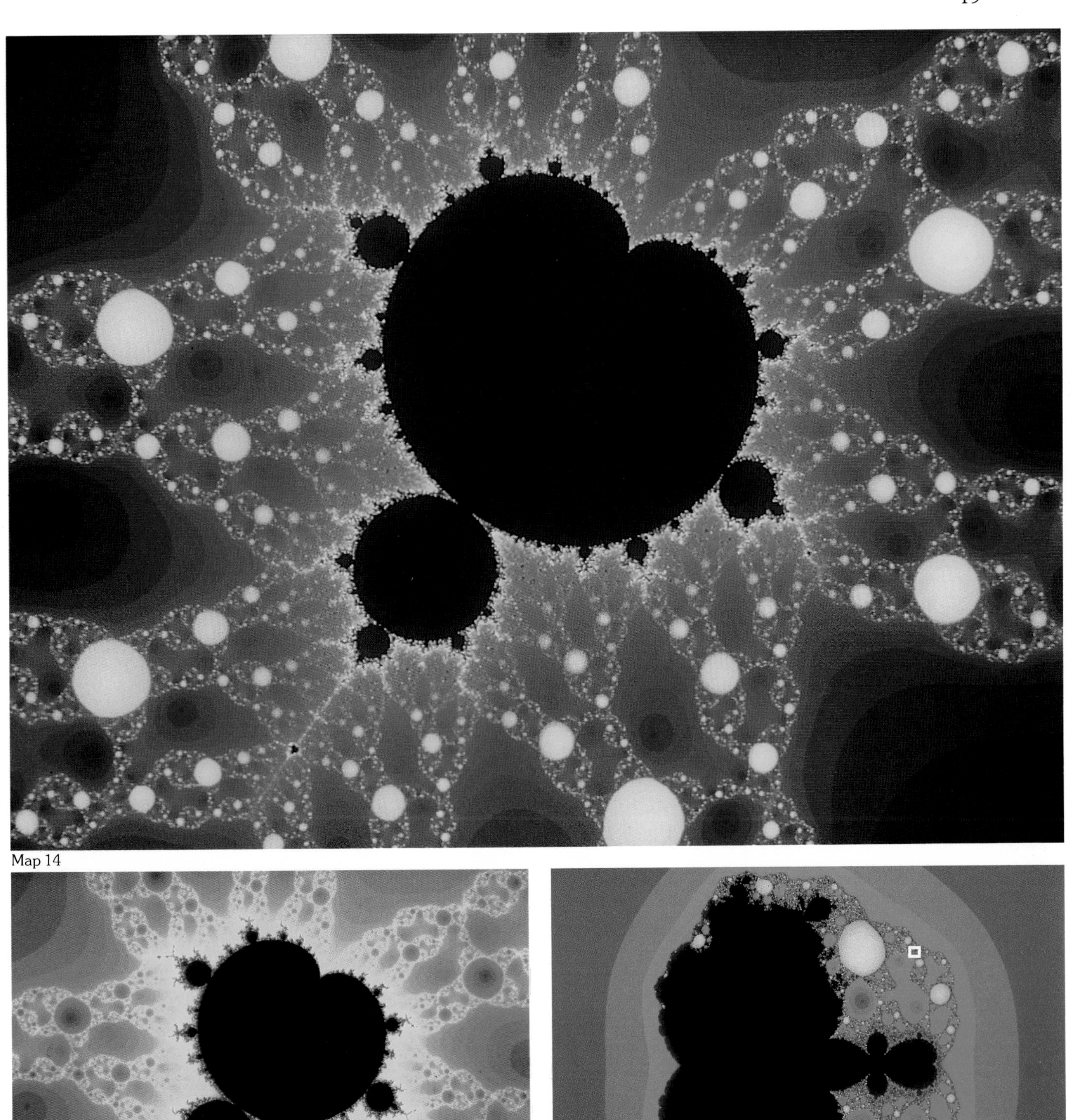

Map 14

Map 15

Map 16

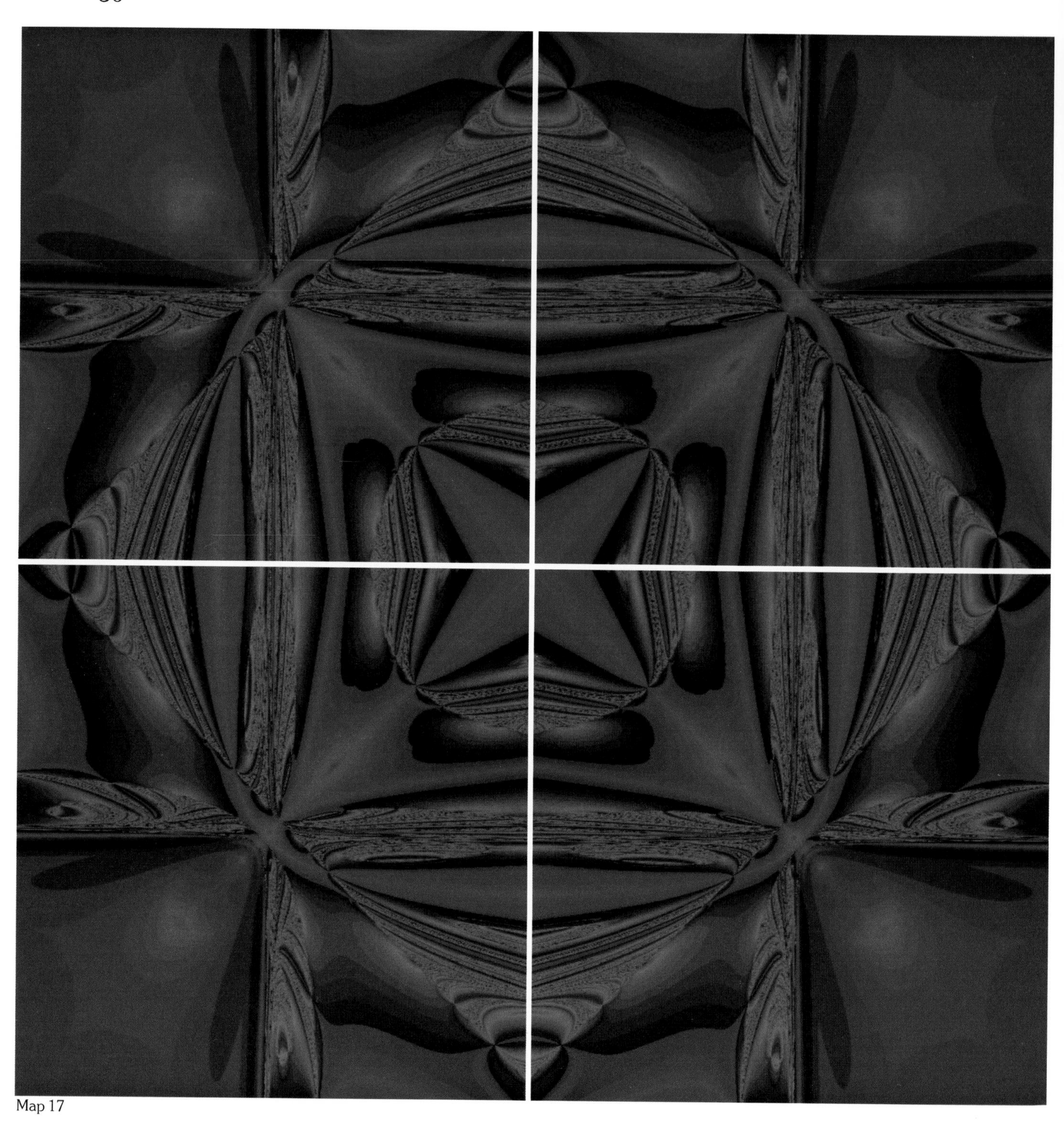

Map 17

Map 18

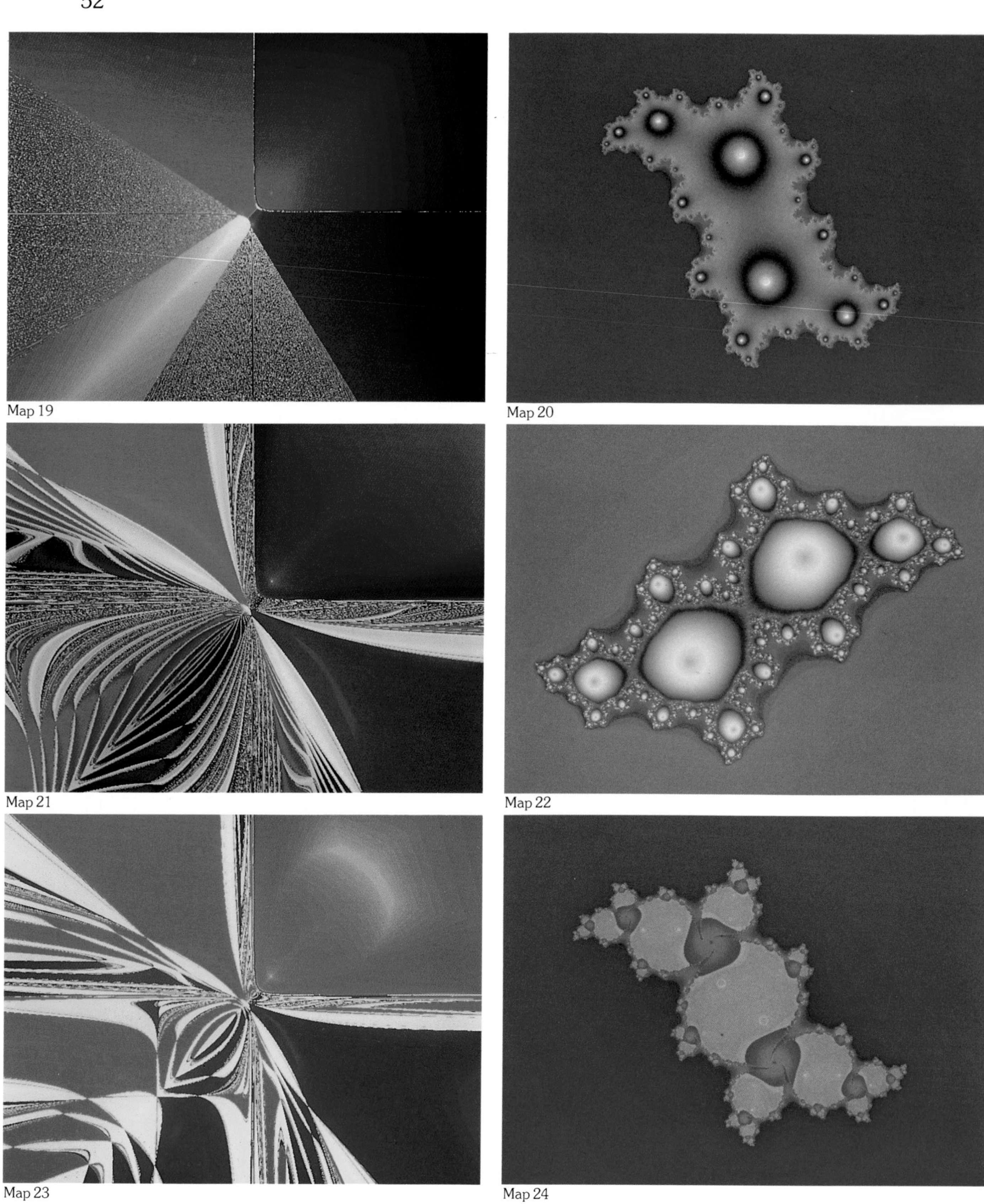

Map 19

Map 20

Map 21

Map 22

Map 23

Map 24

# 3 Sullivan's Classification and Critical Points

From the complexity of the computer experimental results it may appear impossible to understand the *global* dynamics of a given rational map $R$. But Julia and Fatou already knew that many of its qualitative aspects are intimately linked to the dynamics of the critical points of $R$.

(3.1) A number $c \in \overline{\mathbb{C}}$ is called a *critical value* of $R$ if the equation $R(x) - c = 0$ has a degenerate zero, i.e. a zero of multiplicity greater than *1*. Any such zero is called a *critical point*. Finite critical points are obtained as solutions to $R'(x) = 0$.

To report on the remarkable results of D. Sullivan (1983) [Su], we need the concept of an immediate basin of attraction:
Let $x_0$ be an attractive or rationally indifferent fixed point of $R$ and let $A(x_0)$ denote its basin of attraction. Then the *immediate basin of attraction* $A^*(x_0)$ is the connected component of $A(x_0)$ containing $x_0$. In Fig. 3 $A^*(x_0) = A(x_0)$ which is one of the reasons why $J_c$ is in fact a Jordan curve in that example (i.e. a homeomorphic image of the unit circle). In Fig. 6 we see a parabolic fixed point with the shaded region being $A^*(x_0)$. In Fig's. 53 g–j we have typical examples of attractive fixed points ($x_0 = 1$) with $A^*(x_0) \neq A(x_0)$ (see also Map 10). It is a fact that $A^*(x_0)$ is either simply connected (no holes) or has infinite connectivity (infinitely many holes). Maps 7–9 illustrate the latter situation: the white specks designate holes in a blow up of a part of $A^*(x_0)$.
We can also define the immediate basin of attraction of a periodic orbit $\gamma = \{x_0, R(x_0), R^2(x_0), \ldots, R^{n-1}(x_0)\}$ of period $n$ (i.e. $R^n(x_0) = x_0$) which is attractive or rationally indifferent. First let $A^*(x,S)$ denote the immediate basin of a fixed point $x$ for any mapping $S$. Then for periodic orbits of period $n$ we set

$$(3.2) \quad A^*(\gamma) = \bigcup_{k=0}^{n-1} A^*(R^k(x_0), R^n).$$

In Fig. 4 the shaded region illustrates $A^*(\gamma)$ of an attractive 3-cycle. Similarly, if $\gamma = \{x_0, R(x_0), \ldots, R^{n-1}(x_0)\}$ is a periodic orbit of period $n$, which is irrationally indifferent and such that $R^n$ is linearizable in $x_0$, we set

$$(3.3) \quad D(\gamma) = \bigcup_{k=0}^{n-1} R^k(D_0),$$

where $D_0$ is the Siegel disk of $x_0$ for $R^n$. (In case $x_0$ is a fixed point, $\gamma$ is simply the set $\{x_0\}$.)
We can now state Sullivan's results [Su] which completely characterize the Fatou set $F_R$:

### *No Wandering Domains*

(3.4) The Fatou set $F_R = \overline{\mathbb{C}} \setminus J_R$ has countably many connected components. If $X_0$ is such a component, then $X_0$ is eventually periodic, i.e. for some $k$ one has that $R^k(X_0)$ is a periodic component of $F_R$. In short: $R$ has *no wandering domains.*

This answers one of the major problems left open in the work of Julia and Fatou. (Sullivan's proof uses Teichmüller theory and the theory of Fuchsian and Kleinian groups.) Building on fundamental results of Julia and Fatou, a complete classification is thereby possible.

### *Classification into Five Types*

(3.5) Let $X_0$ be a periodic connected component of $F_R$ of period $n$, and

$$\Gamma = \bigcup_{k=0}^{n-1} R^k(X_0)$$

the associated cycle*. Then $\Gamma$ is one of the following:

(A) An immediate basin $A^*(\gamma)$ associated with a superattractive cycle $\gamma$

(B) An immediate basin $A^*(\gamma)$ associated with an attractive cycle $\gamma$.

(C) An immediate basin $A^*(\gamma)$ associated with a parabolic cycle $\gamma$.

(D) A collection of Siegel disks $D(\gamma)$ associated with an irrationally indifferent cycle $\gamma$ [see (3.3)].

(E) A collection of Herman rings

$$H = \bigcup_{k=0}^{n-1} R^k(H_0).$$

The last alternative was discovered a few years ago by M. Herman. It is not associated with a periodic point but it is similar to case $D$: On $H_0$, $R^n$ is analytically equivalent to an irrational rotation of the standard annulus.
Each of the cases (A)–(E) is furthermore characterized by critical points:

### *Detection by Critical Points*

(3.6) In cases (A)–(C) $A^*(\gamma)$ contains at least one critical point. In cases (D) and (E) the boundary of $D(\gamma)$ or $H$ is in the closure of the forward orbit of a critical point (i.e. there is a critical point $x_c$ such that $Or^+(x_c)$ gets arbitrarily close to $\partial D(\gamma)$ or $\partial H$).

Now if $d$ is the degree of the rational function $R$, it is easy to see that $R$ has at most $2d-2$ critical points. (If $R$ is the quotient of two relatively prime polynomials, then the degree of $R$ is the maximum of the degrees of these.) Thus $R$

---

* Note that we use the term cycle for $\Gamma$ although $\Gamma$ may not contain any periodic point (e.g. in case (E)).

can only have finitely many cycles of type (A)–(E). It is not known whether $2d-2$ is an upper bound. It is conjectured that the boundary of $D(\gamma)$ (see (C)) always contains a critical value. In fact M. Herman was able to support this by showing that it is true for $R(x)=z^m+a$, $m=2, 3, \dots$ $a\in\mathbb{C}$.
As a remarkable application of (3.6) we mention the following example from [MSS]: Let $R(x)=((x-2)/x)^2$, then $J_R=\overline{\mathbb{C}}$! To see this, note that the critical points of $R$ are $\{2,0\}$ and that $2\mapsto 0\mapsto\infty\mapsto 1\mapsto 1$ and $R'(1)=-4$. Thus $\overline{\mathbb{C}}\backslash J_R$ must be the empty set, because otherwise (3.5) and (3.6) would apply.

# 4 The Mandelbrot Set

For polynomials of second order, $p(x) = a_2x^2 + a_1x + a_0$, an almost complete classification of the corresponding Julia sets can be given in terms of the Mandelbrot set. First note that $p(x)$ is conjugate to $p_c(z) = z^2 + c$ by means of the coordinate transformation $x \mapsto z = a_2x + a_1/2$, with $c = a_0a_2 + \frac{a_1}{2}\left(1 - \frac{a_1}{2}\right)$. This transformation shifts the finite critical point $x = -a_1/2a_2$ into the origin. It is thus sufficient to study the nature of the Julia sets of $p_c(z)$.
The point $\infty$ is a superattractive fixed point of the mapping $z \mapsto p_c(z)$. The Julia set $J_c$, for given $c \in \mathbb{C}$, can therefore be characterized as $J_c = \partial A(\infty)$. From the theory of Julia and Fatou it follows that $J_c$ is either connected or a Cantor set [Bl]. This distinction is reflected in the definition of the Mandelbrot set:

(4.1) $M = \{c \in \mathbb{C} : J_c \text{ is connected}\}$.

Figures 3, 4, 6–10, 12 and 14 are examples of connected Julia sets whereas Figs. 11, 13 and 15 show Julia sets with Cantor set structure. Among the connected Julia sets there are those which enclose an interior and others, like Fig. 12, which are dendrites without an inner region.
To compute $M$, B.B. Mandelbrot employed the powerful results of Julia and Fatou according to which the main dynamical features of a rational mapping can be inferred from the forward orbits of its critical points (see special Section 3): Any attractive or rationally indifferent cycle has in its domain of attraction at least one critical point. But $p_c(z)$ has only two critical points, $z = 0$ and $\infty$, which are independent of $c$. The point $\infty$ is already an attractive fixed point, so only 0 remains as an interesting critical point to study. By choosing $c = 1$, e.g., we see that there are values of $c$ for which $0 \in A(\infty)$, since $0 \mapsto 1 \mapsto 2 \mapsto 5 \mapsto 26 \mapsto 677 \mapsto \ldots$. In these cases there cannot be another attractor besides $\infty$. On the other hand, as the case $c = 0$ shows, there are also $c$ such that there is another attractor: under $p_0(z) = z^2$, the point $z = 0$ attracts all $z$ with $|z| < 1$, i.e. $J_o = S^1$.
Now according to Julia and Fatou, $J_c$ is connected if and only if $0 \notin A(\infty)$, see [Bl], i.e.

(4.2) $M = \{c \in \mathbb{C} : p_c^k(0) \nrightarrow \infty \text{ as } k \to \infty\}$.

This characterization is very suitable for numerical studies. One chooses a lattice of points $c \in \mathbb{C}$ and tests for every such $c$ whether after $N$ iterations the modulus of the sequence $0 \mapsto c \mapsto c^2 + c \mapsto \ldots$ is still below a given bound $m$. (For Fig. 2 we took $N = 1000$ and $m = 100$.)
A. Douady and J.H. Hubbard [DH1] have found a deep analytic characterization of $M$. They studied the nature of *filled-in Julia sets* $K_c$

(4.3) $K_c = \{z \in \mathbb{C} : p_c^k(z) \nrightarrow \infty \text{ as } k \to \infty\}$, and noticed that for $c \in M$ their complements can be mapped onto the complement of the closed unit disk $\bar{D}$, by means of a conformal mapping $\varphi_c$,

(4.4) $\varphi_c : \bar{\mathbb{C}} \setminus K_c \to \bar{\mathbb{C}} \setminus \bar{D}$.

Remarkably, this mapping can be chosen in such a way that

(4.5) $\varphi_c \circ p_c \circ \varphi_c^{-1} = p_0$.

Note that locally $\varphi_c$ is guaranteed by Boettcher's result (2.34). This identifies $M$ as

(4.6) $M = \{c \in \mathbb{C} : p_c$ on $A(\infty)$ is equivalent to $z \mapsto z^2\}$.

The conjugation (4.5) is even possible for $c \notin M$, but then it does not hold in all of $A(\infty)$. Nevertheless, it can be extended far enough to hold at the point $z = c$, and by setting

(4.7) $\psi(c) := \varphi_c(c)$,

we have a mapping $\psi : \bar{\mathbb{C}} \setminus M \to \bar{\mathbb{C}} \setminus \bar{D}$ which is a conformal isomorphism. In this way Douady and Hubbard demonstrated that

(4.8) $M$ is a *connected* set

(i. e. $M$ is not contained in the union of two disjoint open nonempty sets). It is still unknown, however, whether $M$ is also *locally connected*, i. e. whether any piece $U \cap M$ of $M$ ($U \subset \mathbb{C}$ open) has the property that for any $z \in U \cap M$ there is a neighborhood $V \subset U$, $z \in V$, such that $V \cap M$ is connected. The difficulty is that one cannot draw on properties of $K_c$ because there are $c$ for which $K_c$ is not locally connected. Nevertheless, it is believed that the local connectedness of $M$ does in fact hold. This would have important consequences one of which is discussed in Special Section 5.
Yet another characterization of $M$ has recently been given by F. v. Haeseler [Ha]. Using the coordinate change $z = 1/u$ one first transforms $p_c$ into the rational mapping $R_c(u) = u^2/(1 + cu^2)$. The superattractive fixed point for all $c$ is then $u = 0$, and in a neighborhood of 0 $R_c$ can be conjugated to $R_0$ (only for $c \in M$, of course, can this conjugation be extended to the entire basin of attraction $A(0)$). Let $\Phi_c(u) = u + a_2(c)u^2 + a_3(c)u^3 + \ldots$ be that local conjugation. Then

(4.9) $M = \{c \in \mathbb{C} : |a_k(c)| \leq k,\ k = 2, 3, \ldots\}$.

This bears an intriguing relationship to the Bieberbach conjecture which was recently proved by L. de Branges [Br]. Let

$S = \{f : D \to \mathbb{C} : f(x) = x + a_2 x^2 + \ldots,\ f$ analytic and injective$\}$,

where $D$ is the open unit disk; the functions in $S$ are called *schlicht* functions. The Bieberbach conjecture was:

(4.10) If $f \in S$ then $|a_k| \leq k$, $k = 2, 3, \ldots$.

As a consequence of (4.9) F. v. Haeseler obtained

(4.11) $M \subset \{c \in \mathbb{C} : |c| \leq 2\}$.

Since $c = -2$ belongs to $M$, the estimate could not be better.

Let us now consider $M$ in more detail. A particularly interesting part of $M$ is

(4.12) $M' = \{c \in \mathbb{C} : p_c \text{ has a finite attractive cycle}\}$.

Since each attractor absorbs a critical point, there can be only one such cycle for each $c$. It turns out that $M'$ is an open set with infinitely many connected components. Each component is characterized by the period of the corresponding cycle. The main cardioid, e.g., contains all $c$ for which $p_c$ has a stable fixed point. By computing $\lambda = dp_c/dz$ at the fixed point and imposing the stability condition $|\lambda| < 1$, we find that this comprises the set

(4.13) $M'_1 = \{c \in \mathbb{C} : c = \frac{\lambda}{2}(1 - \frac{\lambda}{2}), |\lambda| < 1\}$.

To characterize the components $W$ of $M'$ further, Douady and Hubbard consider the eigenvalue $\rho_W(c)$ of the attractive cycle that exists for $c \in W$. They show that the mapping

(4.14) $\rho_W : W \to D$

is a conformal isomorphism. Thus each component $W$ has a well-defined *center* $c_W$ whose corresponding attractive cycle is superstable, $\rho_W(c_W) = 0$. Let $\{z_1, z_2, \ldots, z_k\}$ be the attractor for a given $c \in W$. Then $\rho_W(c) = 2^k \prod_i z_i$. If this is to be zero, the critical point $z = 0$ must belong to the cycle. The centers of components with $k$-periodic attractors are therefore given by

(4.15) $p_c^k(0) = 0$.

This equation is of degree $2^{k-1}$ in $c$ so that there may be up to $2^{k-1}$ components with $k$-periodic attractors. We give a list of centers with periods up to 4.

(4.16) $k = 1$: $c = 0$; the corresponding component of $M'$ is $M'_1$ as given in (4.13).

(4.17) $k = 2$: $c^2 + c = 0$ with 2 solutions $c = 0$ and $c = -1$. The center $c = 0$ has already been obtained for $k = 1$. There is thus one component $W = M'_2$ with stable orbits of period 2; it is the disk of radius $1/4$ around $c = -1$.

(4.18) $k = 3$: $(c^2 + c)^2 + c = 0$. Ignoring the solution $c = 0$, it remains to solve $c^3 + 2c^2 + c + 1 = 0$. The real solution $c = -1.7549$ is the center of the secondary Mandelbrot set shown in Map 32 while the two complex solutions $c = -0.1226 \pm 0.7449i$ are the centers of the most prominent buds on $M'_1$.

(4.19) $k = 4$: Two of the eight solutions of $p^4(0) = 0$ have already been obtained with $k = 2$. Of the remaining 6, two are on the real axis: $c = -1.3107$ is the center of the bud that develops from $M'_2$ by period doubling, $c = -1.9408$ is from a satellite near the tip of the main antenna. The four complex solutions are $c = 0.282 \pm 0.530i$, corresponding to buds on $M'_1$, and $c = -0.1565 \pm 1.0323i$ one of which is the center of the Mandelbrot figure in the cover picture of this book.

In addition to the center $c=0$, we have 15 centers of period 5, and so on. Obviously, by going to higher and higher periods we would find centers closer and closer to the boundary of $M$.
The mapping (4.14) can be extended to the boundaries of $W$ and $D$. If $c \in \partial W$ and $\rho_W(c) = \exp(2\pi i\alpha)$, it is said that $c$ is a point of *internal angle* $\alpha$. The point of internal angle $\alpha=0$ is called the *root* of $W$; this is where $W$ buds from another component of $M'$ or, if $W$ is a *primitive* component not budding from another one, the root is the cusp of the cardioid. If $c \in \partial W$ is a point of rational internal angle $\alpha = p/q$, then there is a *satellite* component budding from $W$ at the point $c$, whose attractive cycles have $q$ times the period of the cycles of $W$. For example, for $\lambda = \exp(2\pi i p/q)$ in (4.13) we can identify the points on $M_1$ where satellites are attached. For $p/q = 1/2$ we find $c = -3/4$; the period 3 buds $(p/q = \pm 1/3)$ are attached at $c = (-1 \pm 3\sqrt{3}\,i)/8 = -0.1250 \pm 0.6495i$, and period 4 buds grow from $c = 1/4 \pm i/2$. For all these values of $c$, the mapping $p_c$ has rationally indifferent cycles. Figures 6 and 8 correspond to $p/q = 3/5$ and $1/20$ respectively; for Map 18 we chose $p/q = 1/11$. If, on the other hand, the internal angle is sufficiently irrational in the sense of the diophantine condition (2.16), we find Siegel disks occurring. Fig. 7 and Map 25 derive from $c = \rho_{M_1}^{-1}(\exp\{2\pi i\alpha\})$, see (4.13), with $\alpha = (\sqrt{5}-1)/2$.
So far we have discussed the components of $M'$ with their centers and roots. Another conspicuous feature of $M$ are the tips and the branch points of its antennas. For example, the point $c=-2$ is characterized by the critical point being mapped into the repulsive fixed point, $0 \mapsto -2 \mapsto 2 \mapsto 2$. More generally, for the $c$-values in question, the point $z=0$ is preperiodic but not periodic; it is eventually drawn into a repulsive cycle.

$$(4.20) \quad p_c^n(0) = p_c^{n-k}(0), \; n \geq 3, \; n-2 \geq k \geq 1.$$

Such $c$-values are known as Misiurewicz points.
Consider only the simplest cases $n=3$ and $n=4$:

(4.21) $n=3, k=1$:
$(c^2+c)^2 + c = c^2 + c$. Discarding the solution $c=0$ (for which we know that $z=0$ is itself periodic) we only have $c=-2$, the tip of the main *antenna*.

(4.22) $n=4, k=1$:
$((c^2+c)^2+c)^2 + c = (c^2+c)^2 + c$. Again discarding $c=0$ and also $c=-2$, we obtain three solutions. The real solution $c=-1.54369$ is a "band-merging point" in the analysis of Großmann and Thomae [GT]; the complex solutions $c = -0.22816 \pm 1.11514i$ are perhaps best seen in Map 28, where they are the antenna tips that reach farthest in the imaginary direction.

(4.23) $n=4, k=2$:
$((c^2+c)^2+c)^2 + c = c^2 + c$. Ignoring the solutions $c=0, -1, -2$ which have already been discussed, we are left with $c = \pm i$. These $c$-values mark ends of the side *antennas* at the top and bottom of Map 28.

Fig. 33. Legend see p. 62

Fig. 34. Legend see p. 62

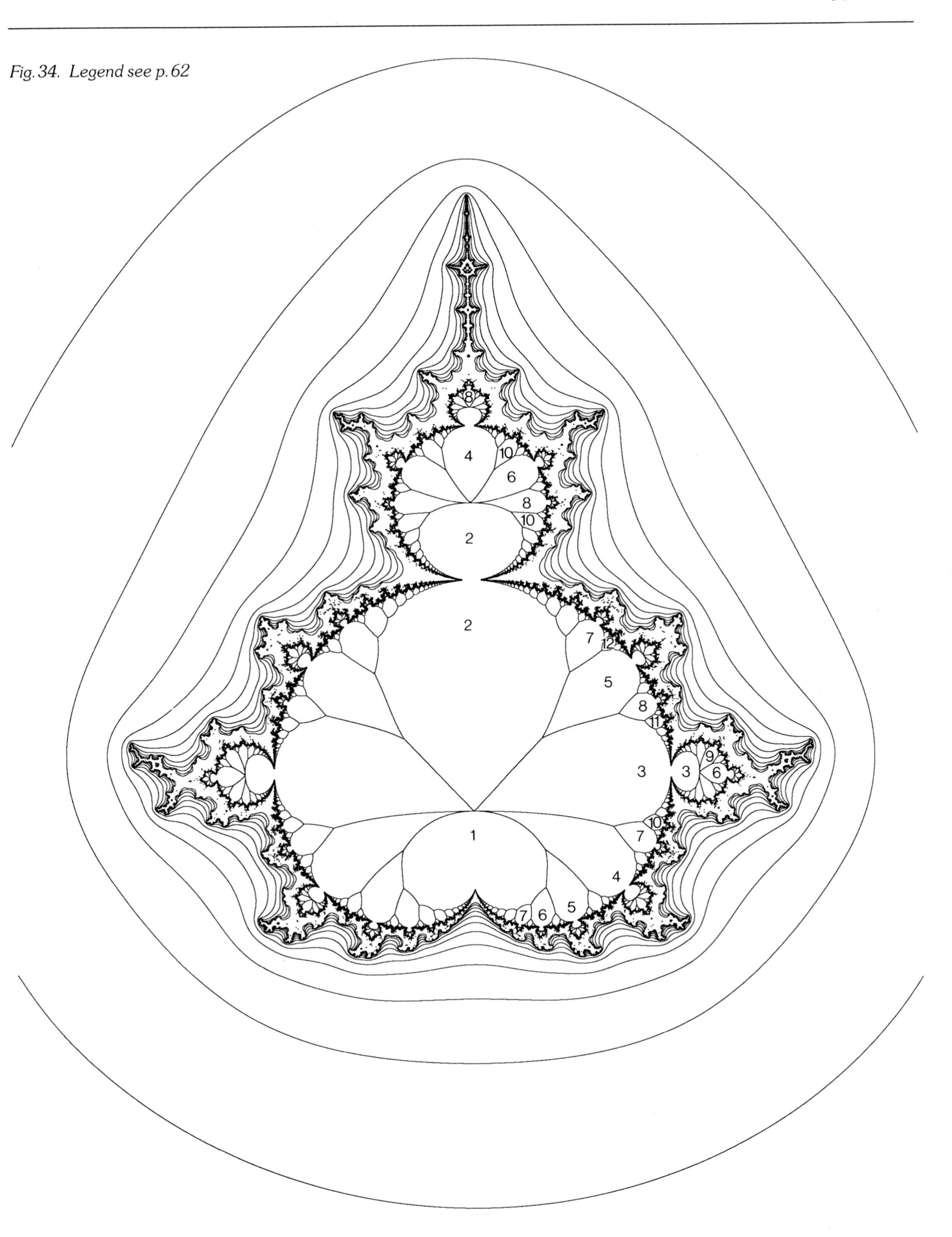

*Fig. 33 (see p. 60). Level sets of $\alpha(c)$ (see (4.24)) in alternating colors for $c \in M$ (based on an experiment by Alfred Barten). Outside of M: equipotential lines*

*Fig. 34 (see p. 61). Domains of index $(c)$ = constant (see (4.24)) for $c \in M$. Indices organize according to Fibonacci sequences. Outside of M: equipotential lines*

The Julia set $J_c$ for $c = i$ is shown in Fig. 12. Its dendrite structure is typical for $c$-values that obey (4.20) and are mapped into repelling cycles: $K_c$ has no interior points, but it is locally connected.
The determination of points $c \in M$ from eqs. (4.15) and (4.20) becomes very cumbersome as we go to higher values of $n$. It is therefore most fortunate that the concept of *external angles* has been developed by Douady and Hubbard. It gives an intuitively very appealing characterization of points $c \in \partial M$, and thanks to a powerful algorithm involving "Hubbard trees" can easily be worked out explicitly. We describe that concept in the Special Section 5.
Figures 33 and 34 are included from some ongoing research on M. Given $c \in M$ we define

$$\alpha(c) = \inf\{|p_c^k(0)| : k = 1, 2, \ldots\}$$

(4.24) and

$$\text{index } (c) = k \text{ provided } \alpha(c) = |p_c^k(0)|.$$

Figure 33 shows levels of $\alpha(c)$ in alternating colors while Fig. 34 shows the distribution of index $(c)$ on $M$. Remarkably each satellite is distinguished by a component of some fixed index and index $(c)$ introduces a Fibonacci partition in $M$.

# 5 External Angles and Hubbard Trees

It is well known that analytic functions $f: \mathbb{C} \to \mathbb{C}$ are a powerful tool for solving problems of two-dimensional electrostatics. The Cauchy-Riemann differential equations imply that $Re f$ and $Im f$ are both solutions to Laplace's equation $\nabla^2 F = 0$, and that the two families of curves $Re f = \text{const}$ and $Im f = \text{const}$ intersect each other orthogonally. Therefore, if $u = Re f$, say, describes the surface of a charged conductor, the lines $Re f = \text{const}$ are *equipotential* lines and $Im f = \text{const}$ the corresponding *field lines*.
Consider the case $z \mapsto w(z) = \ln z$. With $z = |z| \exp(2\pi i \alpha)$ we have $w(z) = \ln|z| + 2\pi i \alpha$. If the unit circle $|z| = 1$ is taken to be a charged conductor, the potential in the surrounding plane is given by $u(z) = Re\, w(z) = \ln|z|$, and the field lines are described by $\alpha = \text{const}$. It is helpful to view $w(z)$ as a conformal mapping which transforms $\mathbb{C} \backslash \bar{D}$ ($D$ being the unit disk) into the half space $Re\, w > 0$, where $Im\, w$ is taken modulo $2\pi$. The important observation is that equipotential lines $|z| = \text{const}$ are mapped into equipotential lines $Re\, w = \text{const}$, and field lines $\alpha = \text{const}$ into field lines $Im\, w = \text{const}$.
Generalizing this observation we may say that a potential problem is solved when a conformal mapping has been found which transforms it to a known problem. Consider, e.g., the problem of a charged line segment

(5.1) $K = \{x \in \mathbb{C}: -2 \leqslant Re\, x \leqslant 2,\ Im\, x = 0\}$.

We are lucky to know a conformal mapping between $\mathbb{C} \backslash K$ and $\mathbb{C} \backslash \bar{D}$ which allows us to carry over what we know about the charged disk:

(5.2) $$\varphi_K: \mathbb{C} \backslash K \to \mathbb{C} \backslash \bar{D}$$
$$x \mapsto z = \frac{1}{2}(x + \sqrt{x^2 - 4})$$

(The branch cut is of course taken to coincide with $K$.) The inverse map is even simpler: $x = \varphi_K^{-1}(z) = z + \frac{1}{z}$. The potential is now easily discussed. Writing

(5.3) $z(x) = \varphi_K(x) = |\varphi_K(x)| \exp(2\pi i \alpha_K(x))$,

we have the potential $u_K(x) = \ln|\varphi_K(x)|$ and field lines $\alpha_K(x) = \text{const}$. The number $\alpha_K(x)$ is called the *external angle* of point $x$. The equipotential lines are the ellipses

(5.4) $$\frac{(Re\ x)^2}{(r + \frac{1}{r})^2} + \frac{(Im\ x)^2}{(r - \frac{1}{r})^2} = 1,\ r > 1,$$

and the field lines are hyperbolas,

(5.5) $$\frac{(Re\ x)^2}{4\cos^2 2\pi\alpha_K} - \frac{(Im\ x)^2}{4\sin^2 2\pi\alpha_K} = 1.$$

Figure 35 shows the relation between $\mathbb{C} \backslash \bar{D}$ and $\mathbb{C} \backslash K$.

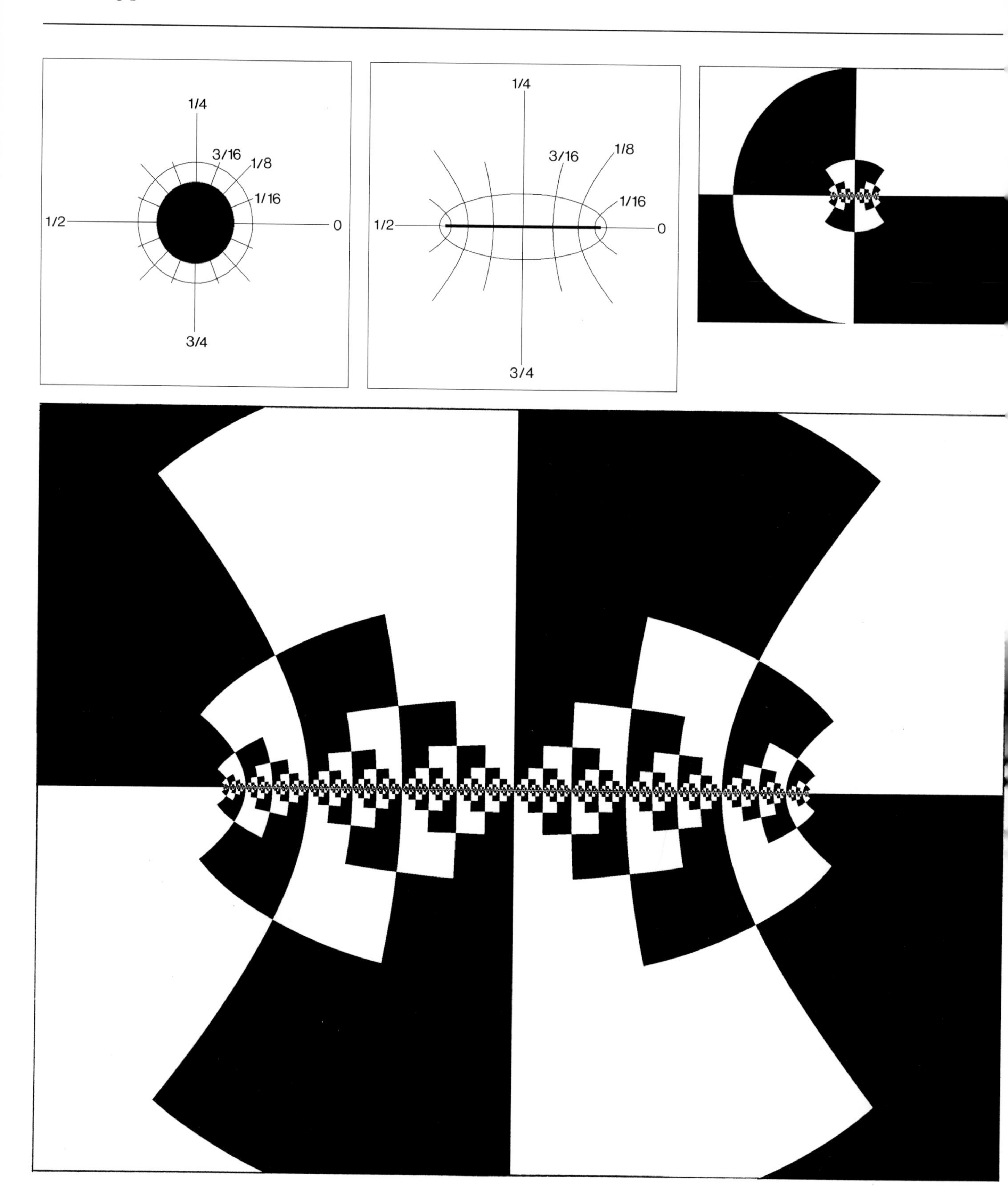
1/4
3/16
1/8
1/16
1/2
0
3/4
1/4
3/16
1/8
1/16
1/2
0
3/4

## The Potential of Connected Julia Sets

Let us now develop the relationship between two-dimensional electrostatics and the process $x \mapsto p_c(x) = x^2 + c$. We begin by noting that the set $K$ in (5.1) is the Julia set for $c = -2$. The remarkable feature of the mapping $\varphi_K$ (5.2) is that it conjugates $p_{-2}$ to $p_0$:

(5.6) $\varphi_K \circ p_{-2} \circ \varphi_K^{-1} = p_0$ on $\overline{\mathbb{C}} \setminus \overline{D}$

A similar conjugation holds for other values of $c$ as well, even though the mapping $\varphi_K$ cannot be given in closed form as in (5.2). This follows very generally from Riemann's theory of conformal mappings, given that the filled-in Julia sets $K_c$ (see 4.3) are compact and their complements $\overline{\mathbb{C}} \setminus K_c$ are connected. Let us first discuss the case $0 \in K_c$ in which $K_c$ is itself connected. The precise statement [DH1–3] is that there is a conformal mapping $\varphi_c : \overline{\mathbb{C}} \setminus K_c \rightarrow \overline{\mathbb{C}} \setminus \overline{D}$ such that

(5.7) $\varphi_c \circ p_c \circ \varphi_c^{-1} = p_0.$

From this general conjugation a surprisingly good approximation can be derived for the *potential* $u_c(x) = \ln|\varphi_c(x)|$. The reason is that on $\overline{\mathbb{C}} \setminus \overline{D}$ the potential rises by a factor of 2 when we apply $p_0$. Correspondingly, for $x \in \overline{\mathbb{C}} \setminus K_c$ we have $u_c(x) = \frac{1}{2} u_c(p_c(x)) = \ldots = \frac{1}{2^n} u_c(p_c^n(x))$, and if $p_c^n(x)$ is sufficiently large the potential approaches that of the disk:

(5.8) $$u_c(x) = \lim_{n \to \infty} \frac{1}{2^n} \ln |p_c^n(x)|.$$

Using $p_c^n(x) = (p_c^{n-1}(x))^2 + c$ and iterating we find

(5.9) $$u_c(x) = \ln |x| + \sum_{n=1}^{\infty} \frac{1}{2^n} \ln \left| 1 + \frac{c}{(p_c^{n-1}(x))^2} \right|.$$

This rapidly converging series can be used to compute equipotential lines around very complicated Julia sets.

What can we say about the *external angles*? Seen from far away, $K_c$ resembles a point charge. The field lines of external angles $\alpha_c$ will therefore approach the rays $r \exp(2\pi i \alpha_c)$ as $r \rightarrow \infty$. As we follow a field line toward $K_c$ the question arises whether a point $x \in \partial K_c$ can be identified on which the field line ends. This turns out to be a very difficult question. According to Carathéodory, the map $\varphi_c$ has a continuous extension to $\partial K_c$ if and only if $K_c$ is locally connected. This is the case for most $c \in M$, but not for all. In general, all that can be said is that field lines with *rational* external angles can be continued to $\partial K_c$. If $\alpha_c = p/q$ with $q$ odd, the field line ends in a repulsive periodic point or in a rationally indifferent point. If $\alpha_c = p/q$ and $q$ even, the endpoint is a pre-

◁ *Fig. 35. Equipotential and field lines for the Julia set of* $x \mapsto x^2 + c$, $c = -2$. Above left and center: *the conjugation* $\varphi_K$; above right and below: *a binary decomposition of* $A(\infty)$ *(white* $\leftrightarrow 0$, *black* $\leftrightarrow 1$*)*

periodic point which is itself nonperiodic. The angle $\alpha_c=0$ always leads to the unstable fixed point of $p_c$, and $\alpha_c=1/2$ to its preimage on the Julia set.

### *Binary Expansion of External Angles*

For rational external angles, the field lines can be continued to $\partial K_c$. Their endpoints are the images, under $\varphi_c^{-1}$, of the rational points $\exp(2\pi ip/q)$ on the unit sphere $S^1$. In view of the conjugation (5.7), these points on $\partial K_c$ behave under iteration of $p_c$ as the corresponding points on $S^1$ behave under iteration of $p_0$. The latter process is, however, well understood. There is a nice way to describe the dynamics of a point $z=\exp(2\pi ip/q)$ under $z\mapsto z^2$:
Let $\alpha=p/q$ and $0<\alpha<1$ and let

$$\alpha=\sum_{k=1}^{\infty} a_k 2^{-k},\ a_k\in\{0,1\}$$

be the binary expansion of $\alpha$. Furthermore let $p$ and $q$ be relatively prime and $q\neq 2^n$, $n\in\mathbb{N}$. Then the sequence of binary digits $a_k$ of $\alpha$ unfolds in the iteration $z\mapsto z^2\mapsto z^4\mapsto\ldots$:

$$a_{k+1}=\begin{cases}0, \text{ if } z^{2^k} \text{ is in the upper half plane}\\ 1, \text{ else.}\end{cases}$$

For example, $p/q=2/7$ develops as

$$2/7\mapsto 4/7\mapsto 1/7\mapsto 2/7\mapsto, \text{ i.e. } 2/7=.\overline{010}\ldots$$

The angle $1/12$ produces

$$1/12\mapsto 2/12\mapsto 4/12\mapsto 8/12\mapsto 4/12, \text{ i.e. } 1/12=.00\overline{01}\ldots$$

Conversely, if a given periodic, preperiodic or rationally indifferent point on $\partial K_c$ can be followed under iteration of $p_c$, and the sequence of zeroes and ones is determined (with respect to the line that connects the points $\alpha_c=0$ and $\alpha_c=1/2$ inside $K_c$), then that sequence is the binary expansion of the corresponding external angle.

### *The Potential of the Mandelbrot Set*

If $c\notin M$ we know that the Julia set $J_c$ decays into a Cantor set. It cannot be expected that the conjugation (5.7) still holds everywhere outside $J_c=K_c$. The electrostatic analogy suggests that there should be infinitely many saddle points in the potential which cannot be transformed away by a conformal mapping. However, in a neighborhood of $\infty$ it should still be possible to con-

*Fig. 36. External angles and binary decomposition of* $A(\infty)$ *in a parabolic situation;* ▷ $x\mapsto x^2+c$, $c=-0.481762-0.531657i$. Above left: *the Julia set;* above right: *binary decomposition of* $A(\infty)$*, total view;* below: *blow up of* above right

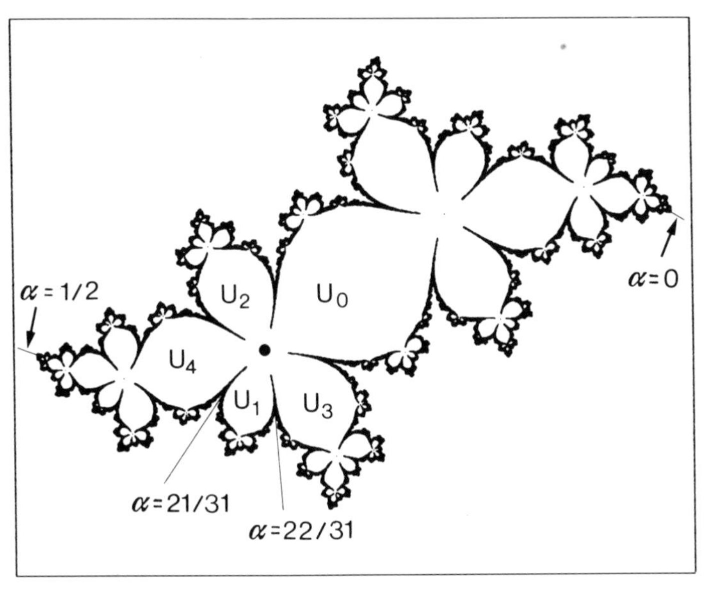

$\alpha = 1/2$
$U_2$
$U_0$
$\alpha = 0$
$U_4$
$U_1$
$U_3$
$\alpha = 21/31$
$\alpha = 22/31$


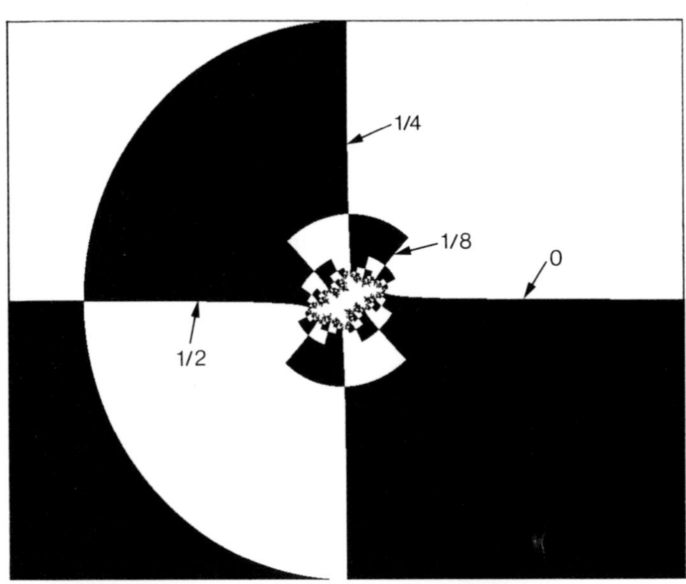

1/4
1/8
0
1/2


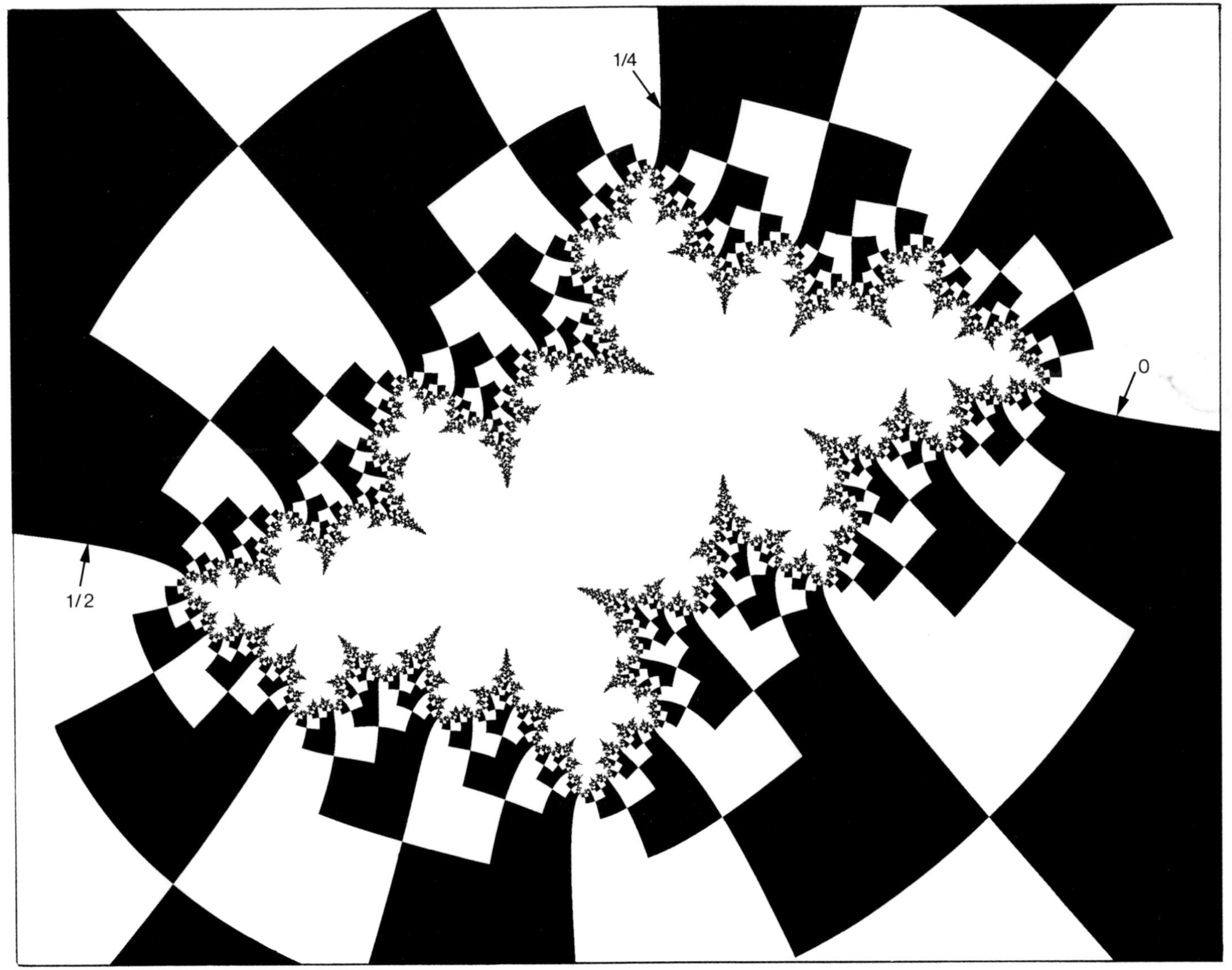

1/4
0
1/2

jugate $p_c$ to $p_0$. Indeed, Douady and Hubbard have shown that this conjugation can be extended sufficiently far to include the point $c = p_c(0)$.
Using the conjugating functions $\varphi_c$, they prove that

$$(5.10) \quad \psi(c) := \varphi_c(c)$$

defines a conformal isomorphism $\overline{\mathbb{C}} \backslash M \to \overline{\mathbb{C}} \backslash \bar{D}$ which solves the potential problem outside the Mandelbrot set:

$$(5.11) \quad U(c) = \ln|\psi(c)| = \ln|c| + \sum_{n=1}^{\infty} \frac{1}{2^n} \ln\left|1 + \frac{c}{(p_c^{n-1}(c))^2}\right|.$$

This formula was used in the computation of Fig. 16.
Concerning the corresponding field lines, it is not known whether they can generally be continued to the boundary of $M$. In other words, it is not known whether $M$ is locally connected. For rational external angles, however, the field lines do extend to $\partial M$, so the problem arises how to parametrize the rational part of $\partial M$ in terms of these external angles.
There are two kinds of points of $\partial M$ for which the external angles can so far be determined. On the one hand, there are $c$-values (4.20) such that 0 is preperiodic, but not periodic. It can be shown that the corresponding $K_c$'s are locally connected and that $c \in K_c$ has a finite number of external angles $p/q$, with $q$ even. The external angles of such $c \in \partial M$ are those same numbers, according to (5.10).
The other set of $c$-values on $\partial M$ which can be assigned rational external angles are the roots of the components of $M'$, see Section 4. They have two external angles each, with odd denominator. To determine these angles we use another theorem of Douady and Hubbard. Consider the connected components $U_0, U_1, \ldots, U_{k-1}$ of the interior of $K_c$ that contain $0, p_c(0), \ldots, p_c^{k-1}(0)$ respectively, $k$ being the smallest number such that $U_k = U_0$. The mapping $p_c^k$ extended to $\partial U_i$ has a fixed point $\rho_i$ which is called the *root* of $U_i$. Consider $\rho_1$ and the two external angles $\alpha_1, \alpha_2$ which reach $\rho_1$ alongside $U_1$. These are the two external angles of $c \in \partial M$. Figure 36 illustrates this procedure for $c = -0.481762 - 0.531657i$, which corresponds to the parabolic case of Fig. 6. From the arrangement of the $U_i$ and the location of the external angles $\alpha = 0$, $\alpha = 1/2$, the angle $\alpha_1$ is seen to have the binary expansion $\alpha_1 = .\overline{10101}\ldots = 21/31$ while $\alpha_2 = .\overline{10110}\ldots = 22/31$.
Figure 37 shows the binary decomposition of $\mathbb{C} \backslash K_c$ according to (2.37) for three different values of c. We took $\alpha_0 = 0$ and replaced $\frac{1}{2\pi}$ arg $R^k(x)$ by $\frac{1}{2\pi}$ arg $p_c^k(x)$; the target set is $L_0(\infty) = \{x : |x| \geq \varepsilon^{-1}\}$.

*Fig. 37. Binary decompositions of $A(\infty)$ for $x \mapsto x^2 + c$;* insert: $c = 0$; above: ▷ $c = 0.32 + 0.043i$; below: $c = i$

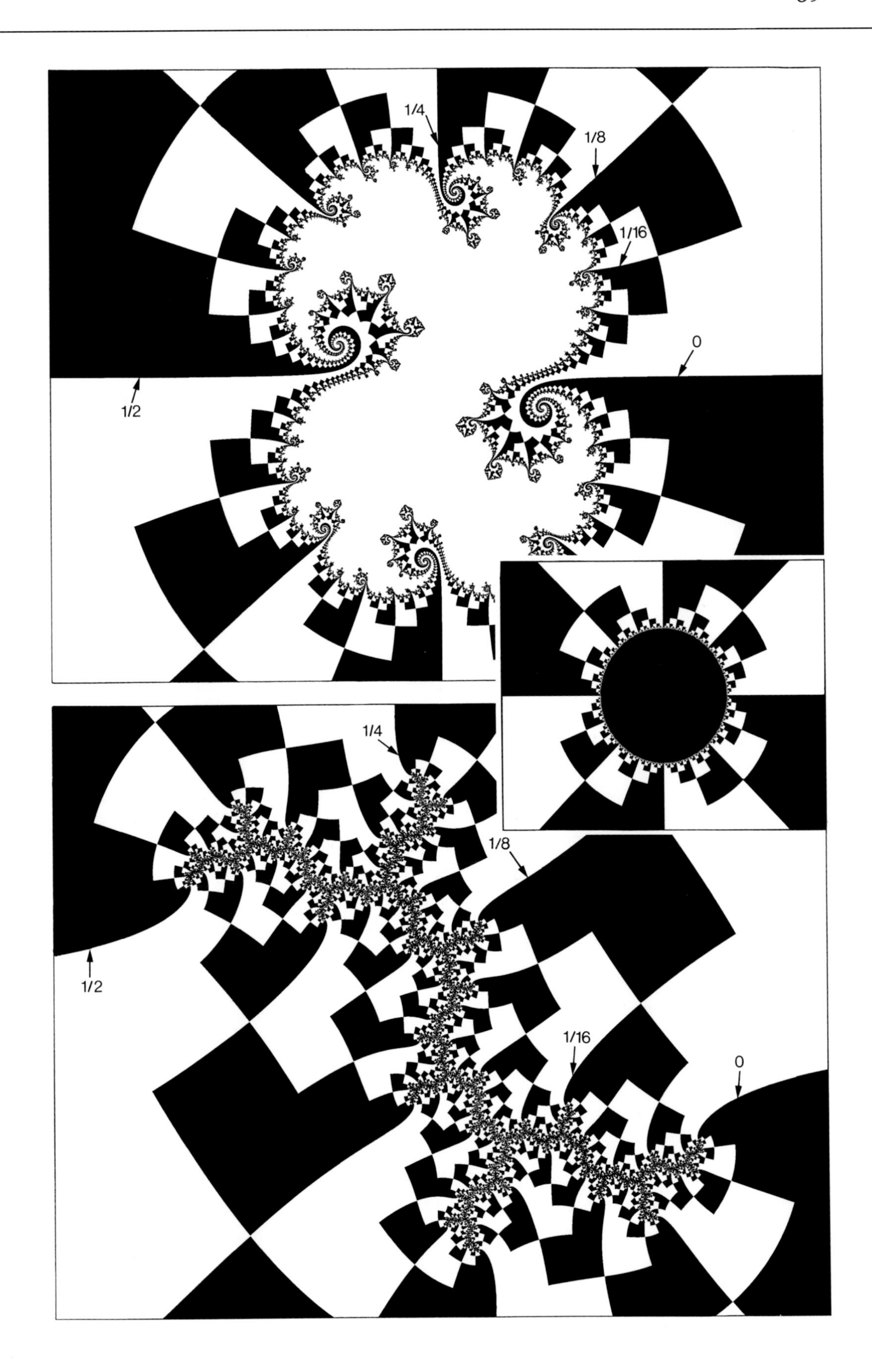
1/4
1/8
1/16
0
1/2
1/4
1/8
1/2
1/16
0

*Hubbard Trees for Preperiodic Critical Points*

J.H.Hubbard devised a rather simple algorithm to calculate external angles of $M$. It only uses the knowledge of the orbits of $x_0 = 0$, for a special choice of $c$-values from $M$. Let us begin to discuss $c \in M$ such that 0 is preperiodic but not periodic. The simplest case (4.21) is $c = -2$. With $x_0 = 0$ we have $x_1 = -2$ and $x_2 = x_3 = 2$. This will be symbolized as

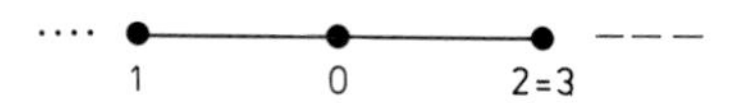


the points marking the orbit of $x_0$, the solid lines being a minimal connection inside $K_c$. The broken and the dotted lines portray external angles 0 and 1/2 respectively for reference. According to (5.10) we need to determine the angle(s) at the point $x_1$ in this diagram (assume we do not know that it is 1/2). To do so, note that application of $p_c$ to the diagram maps the piece (0,1) onto (1,2), expanding it and reversing the orientation, while (0,2) is mapped onto (1,3) which is also expanding but retains the orientation. Now start at $x_1$ above the reference line and keep track of the dynamics in terms of the binary coding: $.0\overline{1}\ldots = 1/2$. Start below the reference line: $.1\overline{0}\ldots = 1/2$. So $c = -2$ admits only one external angle, $\alpha_{-2} = 1/2$.
The next simplest case is $c = -1.54369$, see (4.22). The Hubbard diagram for the orbit of 0 is

Under $p_c$, the piece (0,1) is again expanded and orientation is reversed, whereas (0,2) is simply shifted to (1,3). This time the two possible starts at $x_1$ reveal different angles: $.01\overline{10}\ldots = 5/12$, $.10\overline{01}\ldots = 7/12$. (To understand the period 2 inspite of the fixed point $x_3 = x_4$, remember that the piece (0,1) is turned upside down under the action of $p_c$!) Thus the main band merging point of Großmann and Thomae has external angles 5/12 and 7/12.
Let us also consider the other cases of (4.22). For $c = -0.22816 + 1.11514i$, the Hubbard diagram is

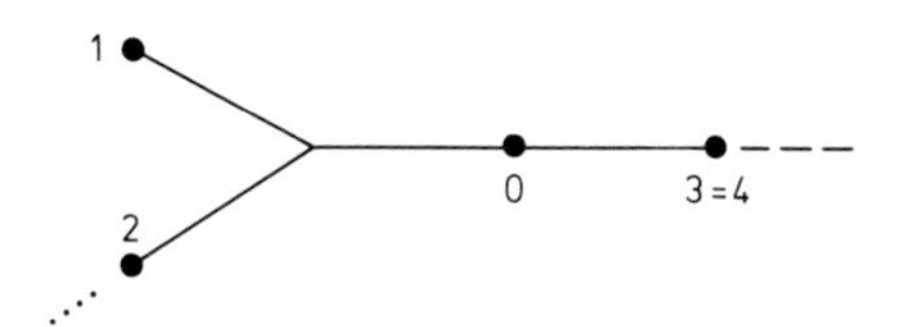


We begin to see why these diagrams are called "trees." The branching reflects the structure of the Julia set. It is a general property of Hubbard trees $H$ that they consist of two parts $H'$ and $H''$ such that $H = H' \cup H''$, $H' \cap H'' = \{x_0\}$, and that the mappings $H' \to H$ and $H'' \to H$ induced by $p_c$ are injective. The location of the dotted line is determined by $x_2$ being the preimage of $x_3$. Starting at $x_1$, we are always above the reference line, but there are two ways to arrive at $x_2$: $.00\overline{1}\ldots = 1/4$, $.01\overline{0}\ldots = 1/4$. Both lead to the same external angle 1/4 for the tip of the antenna in question.

The case $c=i$, see (4.23), is the following:

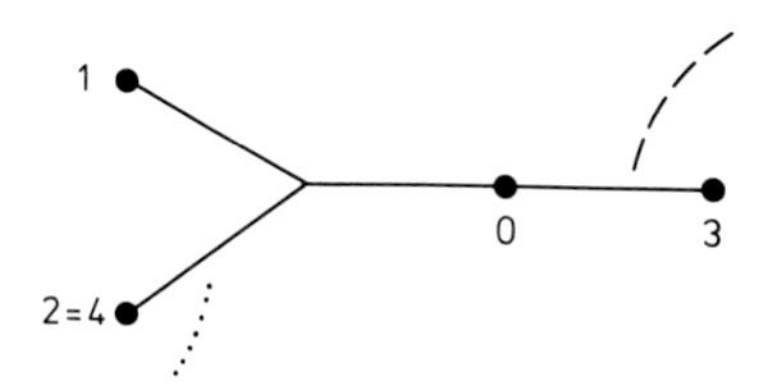


The location of the reference lines deserves a comment. In general, the direction of external angle $\alpha_c=0$ can be obtained from looking at the preimage of the segment (0,1). In the present case, (0,3) is the preimage of (1,4), so the preimage of (0,1) must branch off (0,3) as indicated. The dotted line is the preimage of the broken line. There is only one possible angle: $.0\overline{01}\ldots=1/6$.
As a last example of $c\in\partial M$ such that 0 is preperiodic but not periodic, consider the point where the two last mentioned antennas branch off a common stem. The Hubbard tree is

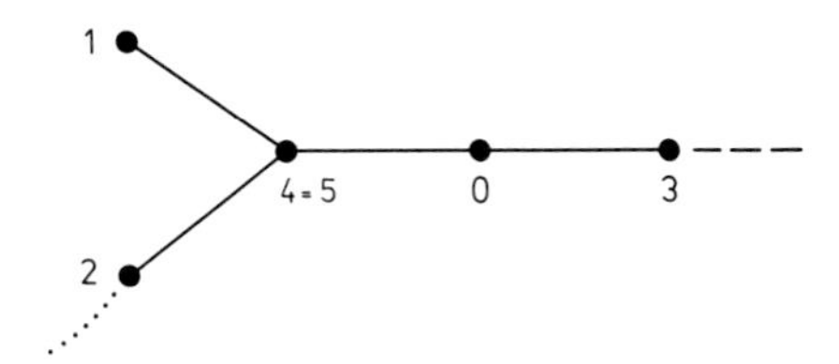


A new complication arises. Starting the process at $x_1$ such that the first three steps are .001, an ambiguity exists at the fourth step. Both choices 0 and 1 are possible. Then following the rotation around the fixed point we have $.001\overline{010}\ldots=9/56$ and $.001\overline{100}\ldots=11/56$. The other possible start does not run into the ambiguity: $.010\overline{001}\ldots=15/56$. As expected, there are three external angles associated with the bifurcation of an antenna.

### *Hubbard Trees for Periodic Critical Points*

If $c\in\partial M$ is the root of a satellite $W$, the Julia set is of the rationally indifferent type. We explained with the help of Fig. 36 how the external angles are then determined. As $c$ moves from the root toward the center $c_W$ of $W$, the parabolic point $\rho_1$ turns into a repulsive point, but its external angles remain the same. Therefore it makes sense to consider $c_W$ rather than the root of $W$, and to use the fact that then the orbit of $x_0=0$ is periodic. Hence to compute the angles of the root $c=-3/4$, we go to the corresponding center $c=-1$ and draw the Hubbard diagram

.... 1 —— 0=2 ---

The two associated angles are obtained by starting at $x_1$, as usual, with 0 or 1: $.\overline{01}\ldots=1/3$ and $.\overline{10}\ldots=2/3$.

Next we discuss other satellites of the main body $M_1'$. The bud of period 3 is represented as

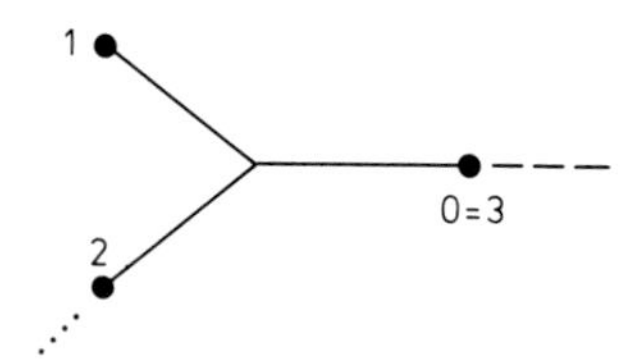


which produces the angles $.\overline{001}\ldots = 1/7$ and $.\overline{010}\ldots = 2/7$. To the right of that bud there is a series of buds of decreasing size but increasing period $k$. Their Hubbard trees are

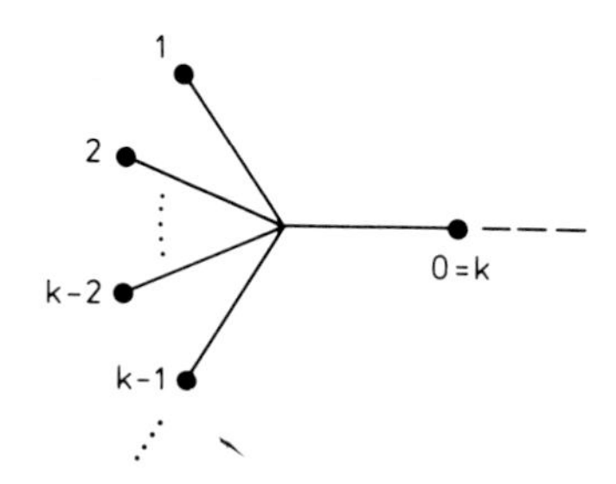


The corresponding angles are $1/(2^k-1)$ and $2/(2^k-1)$. In addition, there are many other series of satellites rooting on $M_1'$. For example, between the two main buds of period 2 and period 3, going down into the "seahorse valley", there is a series with periods 5, 7, 9, .... Their Hubbard trees are

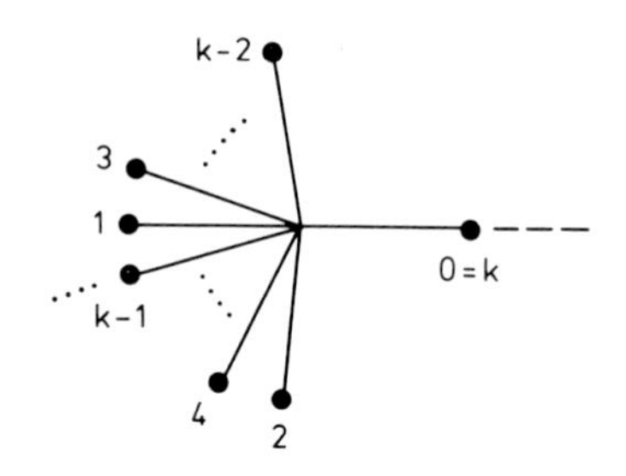


yielding the angles $\frac{1}{3}\,\frac{2^k-5}{2^k-1}$ and $\frac{1}{3}\,\frac{2^k-2}{2^k-1}$.

The secondary Mandelbrot set whose center is at $c = -1.7549$, see (4.18), is associated with

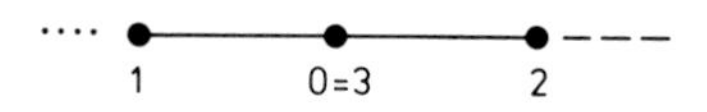

Its two external angles are $.\overline{011}\ldots = 3/7$ and $.\overline{100}\ldots = 4/7$. The center of the Mandelbrot figure on the cover of this book has the Hubbard tree

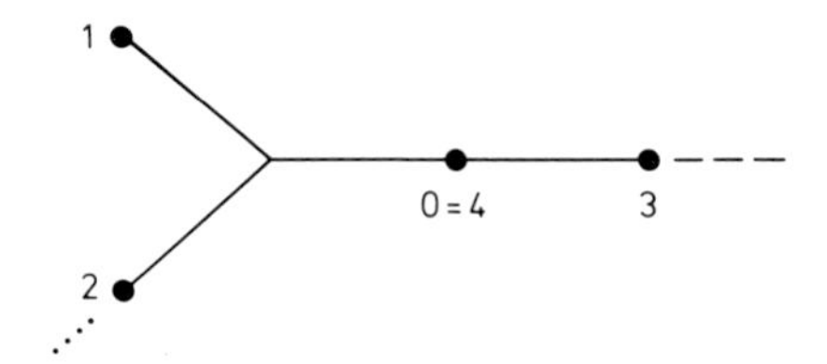


which gives $.\overline{0011}\ldots = 1/5$ and $.\overline{0100}\ldots = 4/15$.
It is entertaining to analyze more and more of these diagrams and to produce Figs. 63 and 64 which summarize some of the results. Analyzed in this way, the boundary of the Mandelbrot set becomes a most interesting representation of the real numbers $\alpha$, $0 \leq \alpha < 1$.

## *Irrational Angles*

The only irrational angles that have so far been shown to allow a continuation to $\partial M$ are associated with period doubling scenarios. The main sequence of period doublings occurs around $c_D = -1.401155$ on the real $c$-axis. To the right of this value, as shown in Fig. 5, there is a sequence of roots whose upper external angles are

$.\overline{01}\ldots = 1/3$
$.\overline{0110}\ldots = 2/5$
$.\overline{01101001}\ldots = 7/17$
$.\overline{0110100110010110}\ldots = 106/257$

The regularity in the binary expansion is obvious. In the limit of infinite bifurcation, the angle approaches the well known Morse-Thue number. To the left of $c_D$, Großmann and Thomae detected the band merging scenario whose $n$-th step is characterized by $c$ entering a repulsive cycle of period $2^{n-1}$ after $2^n$ iterations of $p_c$. The corresponding upper external angles are

$.01\overline{10}\ldots = 5/12$
$.0110\overline{1001}\ldots = 33/80$
$.01101001\overline{10010110}\ldots = 1795/4352$

This sequence converges to the same Morse-Thue number which shows that the corresponding irrational field line has $c_D$ as its limiting point on $M$.
Similarly, the irrational angles of other accumulation points of bifurcations can be determined using Hubbard's diagrams. The bifurcation sequence of the period 3 satellite on the main cardioid gives angles

$.\overline{001} = 1/7$ and $.\overline{010}\ldots = 2/7$
$.\overline{001010}\ldots = 10/63$ and $.\overline{010001}\ldots = 17/63$
$.\overline{001010010001}\ldots$ and $.\overline{010001001010}\ldots$

The regularity is similar to that in the main bifurcation sequence, except for the extra zeros at every third position.

*Fig. 38. Binary decomposition of* $\overline{\mathbb{C}}\backslash M$

*Page 74:*
Above left and below: *total views*
above right: *close up, window as in Maps 52–54*

*Page 75:*
Above: *Binary decomposition of the unit disk in* $\mathbb{C}$
Below: *Representation of* $\overline{\mathbb{C}}\backslash M$ *in the* $1/c$*-plane, total view*

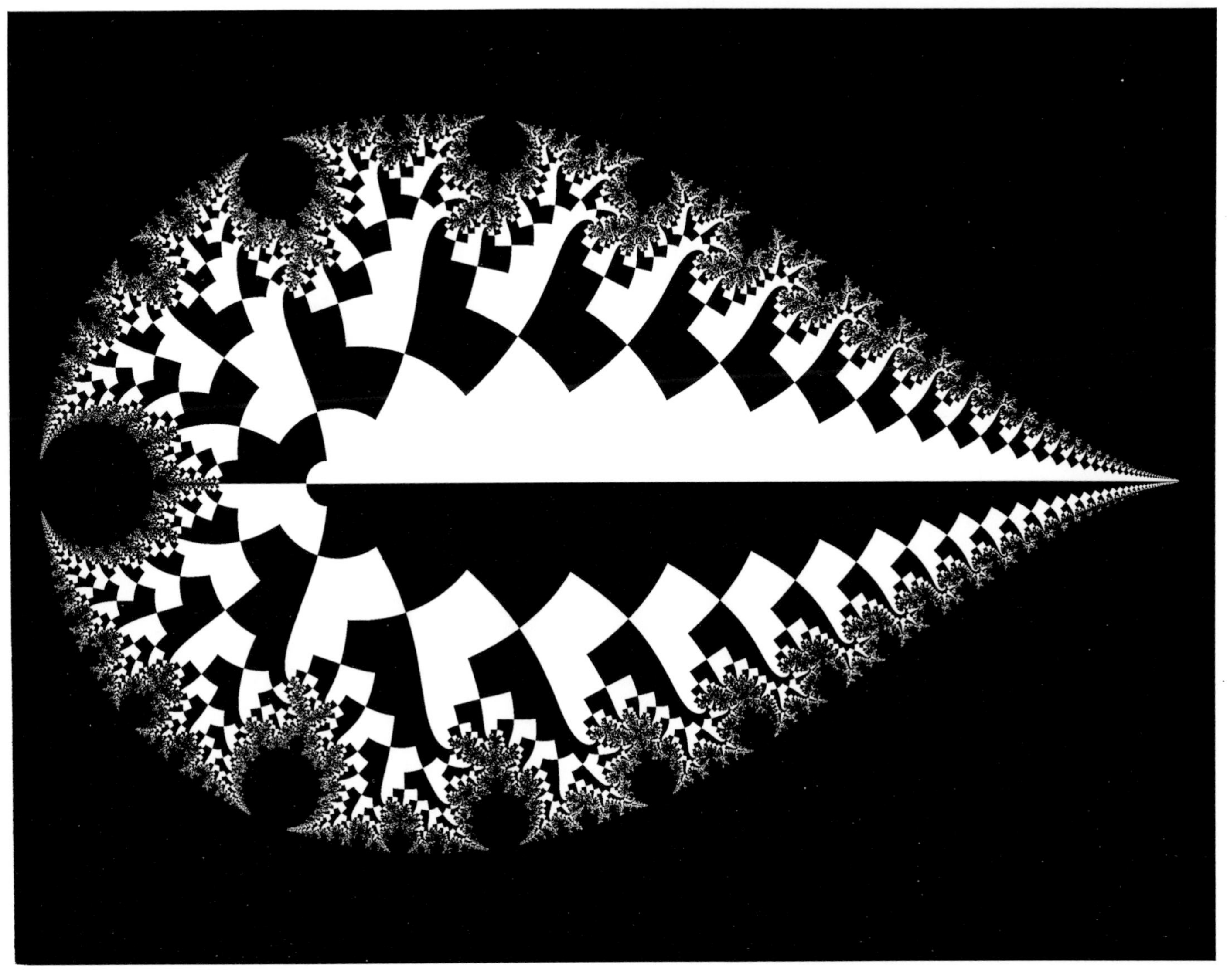

The secondary Mandelbrot set at $c=-1.755$ undergoes bifurcations at upper angles

$.\overline{011}\ldots=3/7$
$.\overline{011100}\ldots=4/9$
$.\overline{011100100011}\ldots=29/65$

Again the regularity allows the definition of the limiting irrational angle.

*Binary Decomposition of* $\overline{\mathbb{C}}\backslash M$

Using techniques strictly analogous to (2.37) we obtained Fig. 38. More precisely, let $L_0(\infty)=\{c\in\mathbb{C}:|c|\geq\varepsilon^{-1}\}$ for $0<\varepsilon\ll 1$, and let $L_k(\infty)$ be the level sets of $M$ introduced in Section 2, i.e.,

$$L_k(\infty)=\{c\in\mathbb{C}:p_c^k(0)\in L_0(\infty) \text{ and } p_c^l(0)\notin L_0(\infty),\, l<k\},\ k=1,2,\ldots$$

Then for $c\in L_k(\infty)$ one determines a coding according to the following scheme:

$$c \text{ is coded } \begin{cases} 0, \text{ if } 0\leq\dfrac{1}{2\pi}\arg(p_c^k(0))\leq\dfrac{1}{2} \\ 1, \text{ else.} \end{cases}$$

($0\leftrightarrow$white, $1\leftrightarrow$black). External angles of the form $p/2^n$ are clearly visible and terminate at the tips of the dendrites in $M$ ($\alpha=1/16, 1/8, 1/4, 5/16, 3/8$ and $1/2$, for example). Also some special values like $\alpha=1/3$ or $\alpha=1/7$ and $\alpha=2/7$ are immediately read off. Figure 38 (p. 75, below) shows the Mandelbrot set and a binary decomposition in the $1/c$-plane, i.e., $L_0(\infty)$ appears now as a small disk centered at the origin, while the Mandelbrot set (black) extends to infinity (in $1/c$-coordinates). This representation is more convenient, because there one can even read off the first binary digits.

Map 25

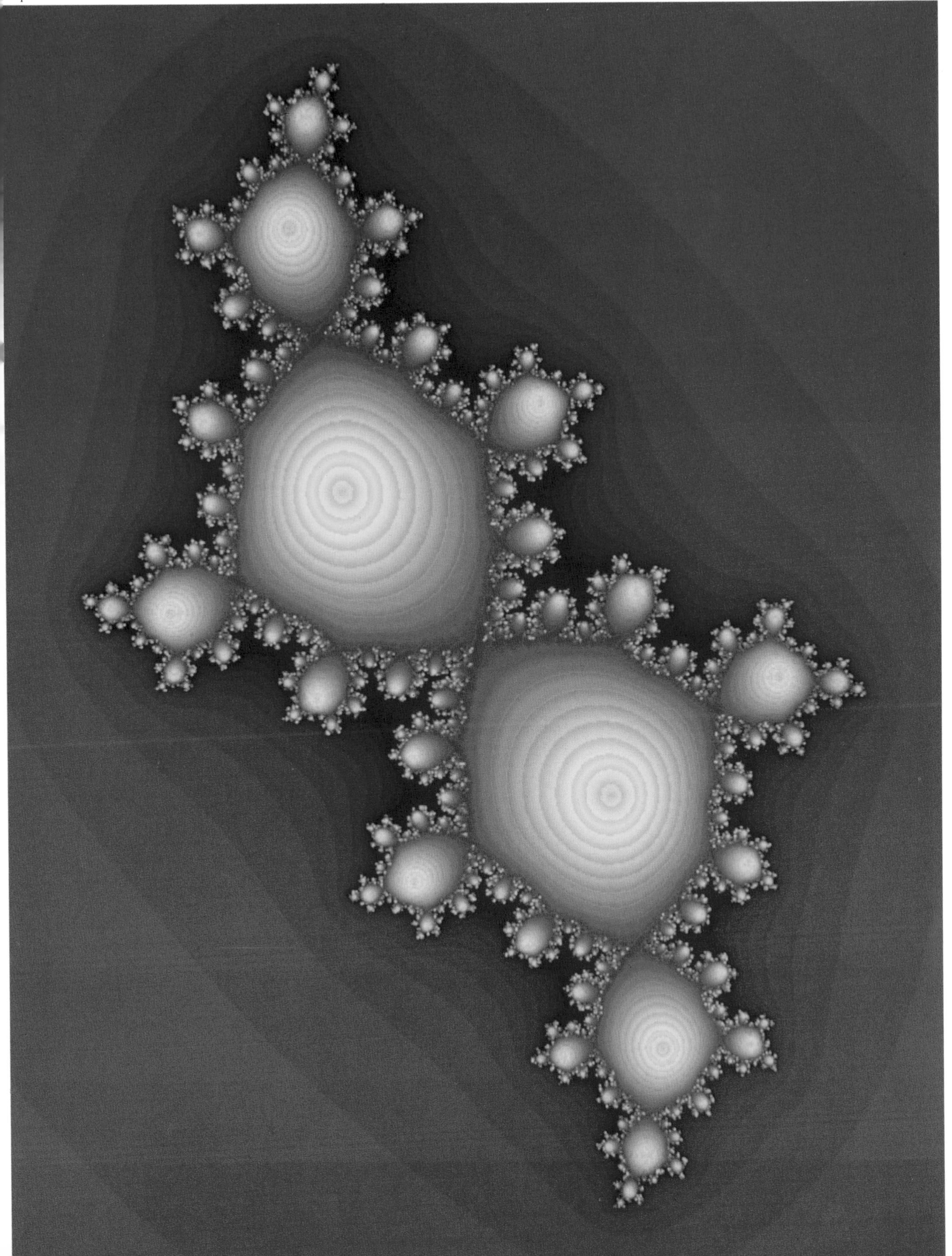

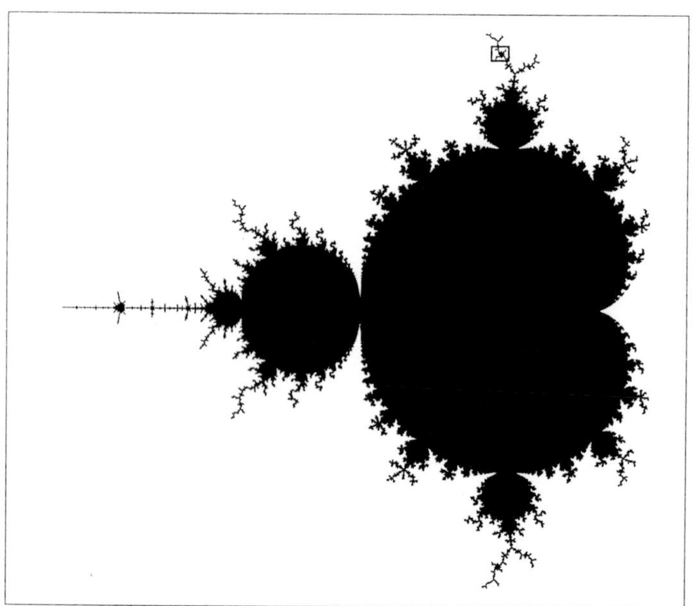

Map 26

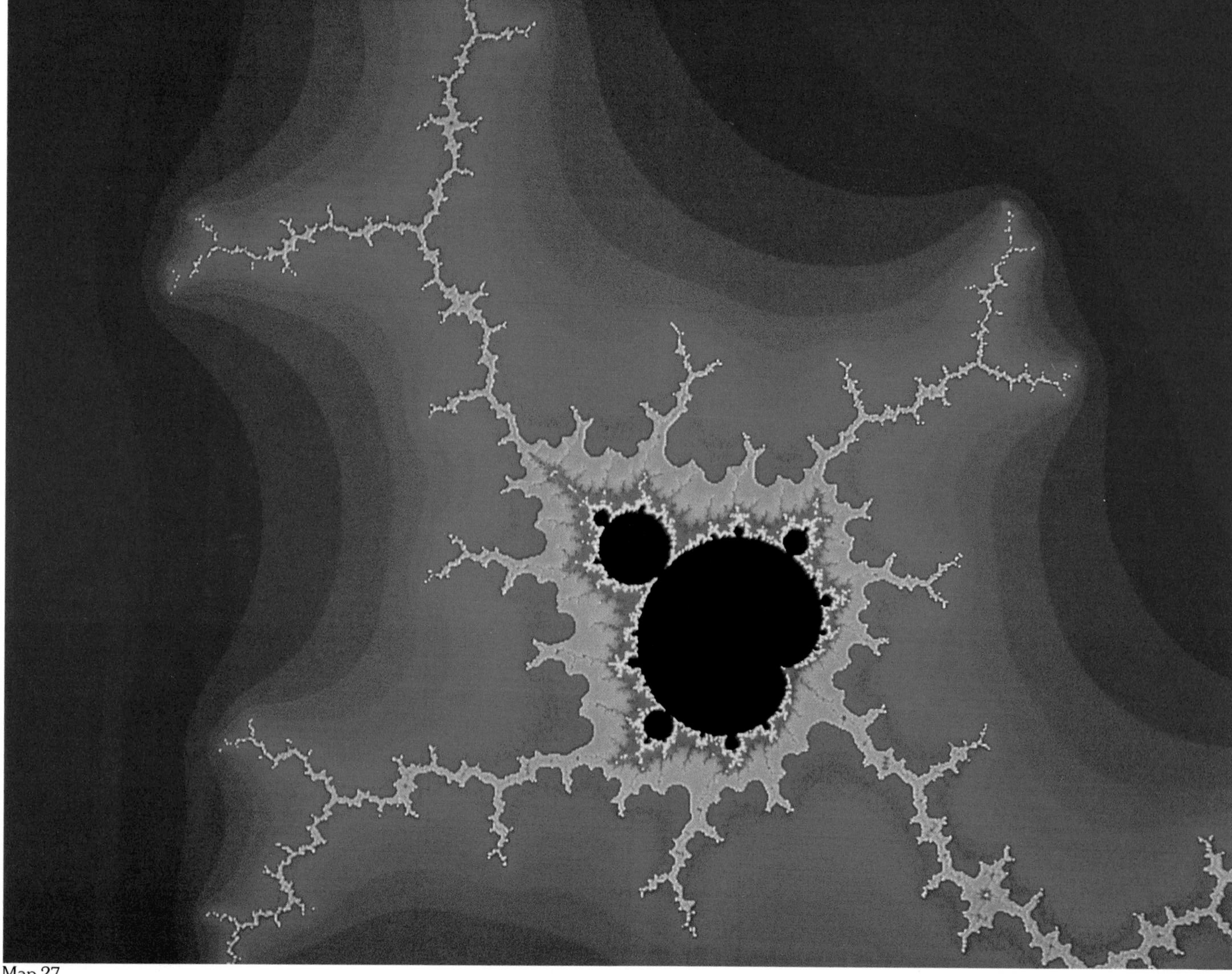

Map 27

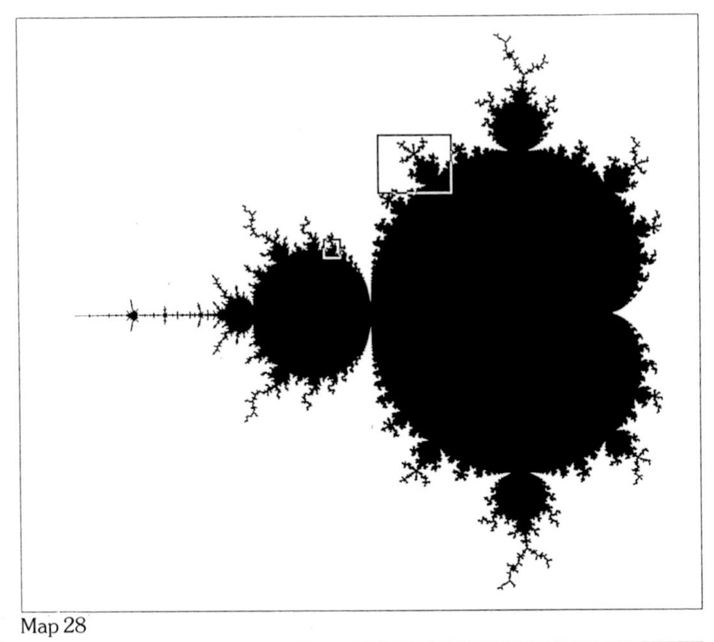

Map 28

Map 29

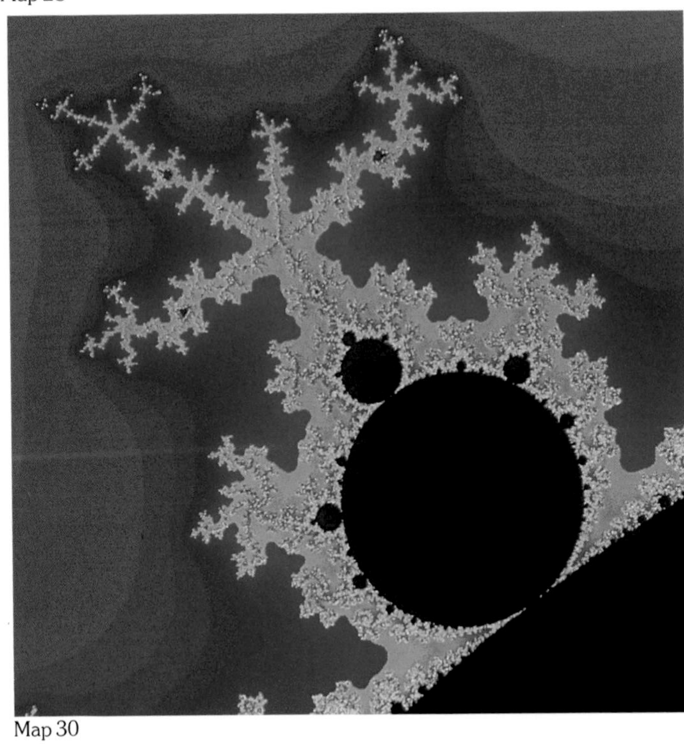

Map 30

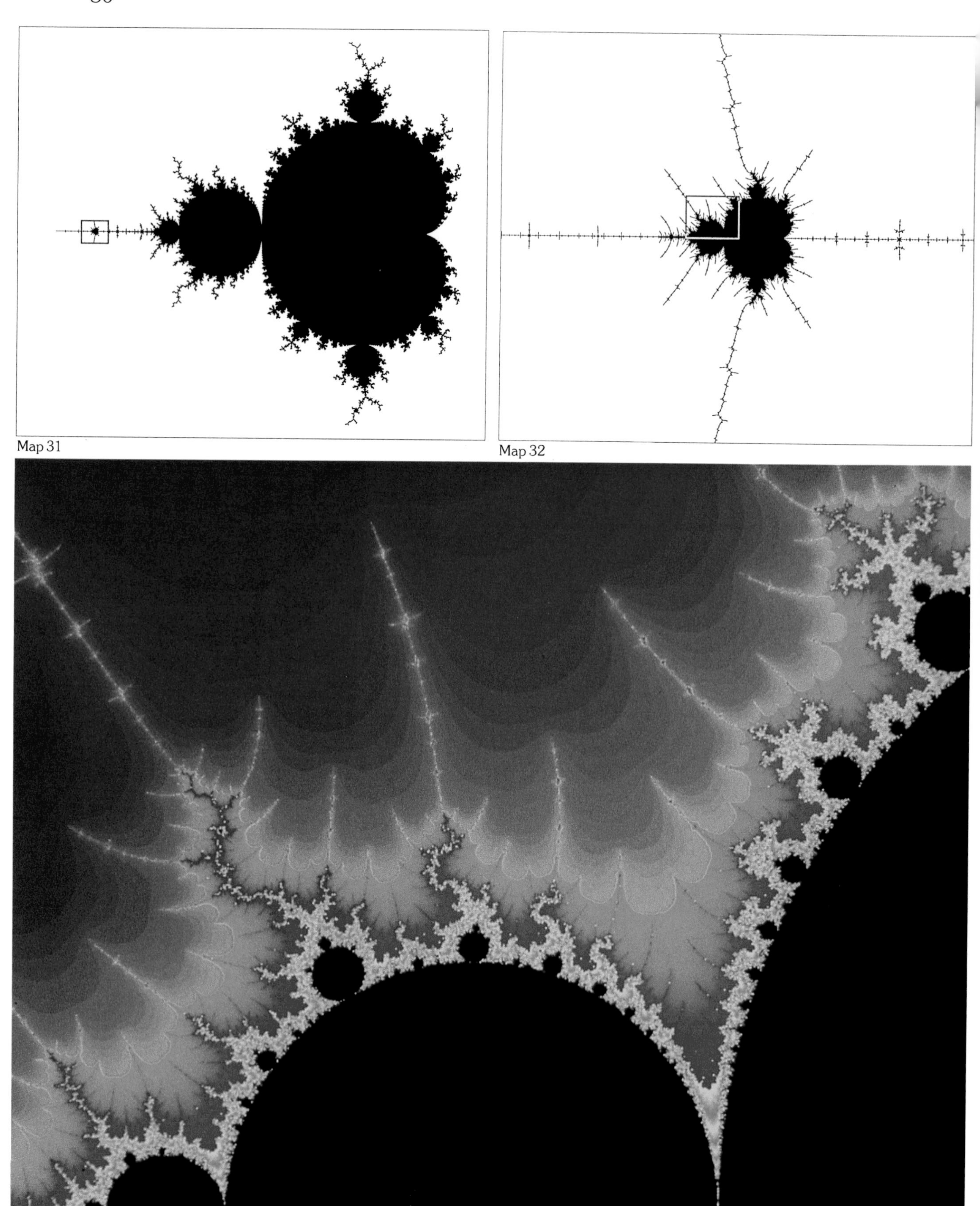

Map 31

Map 32

Map 33

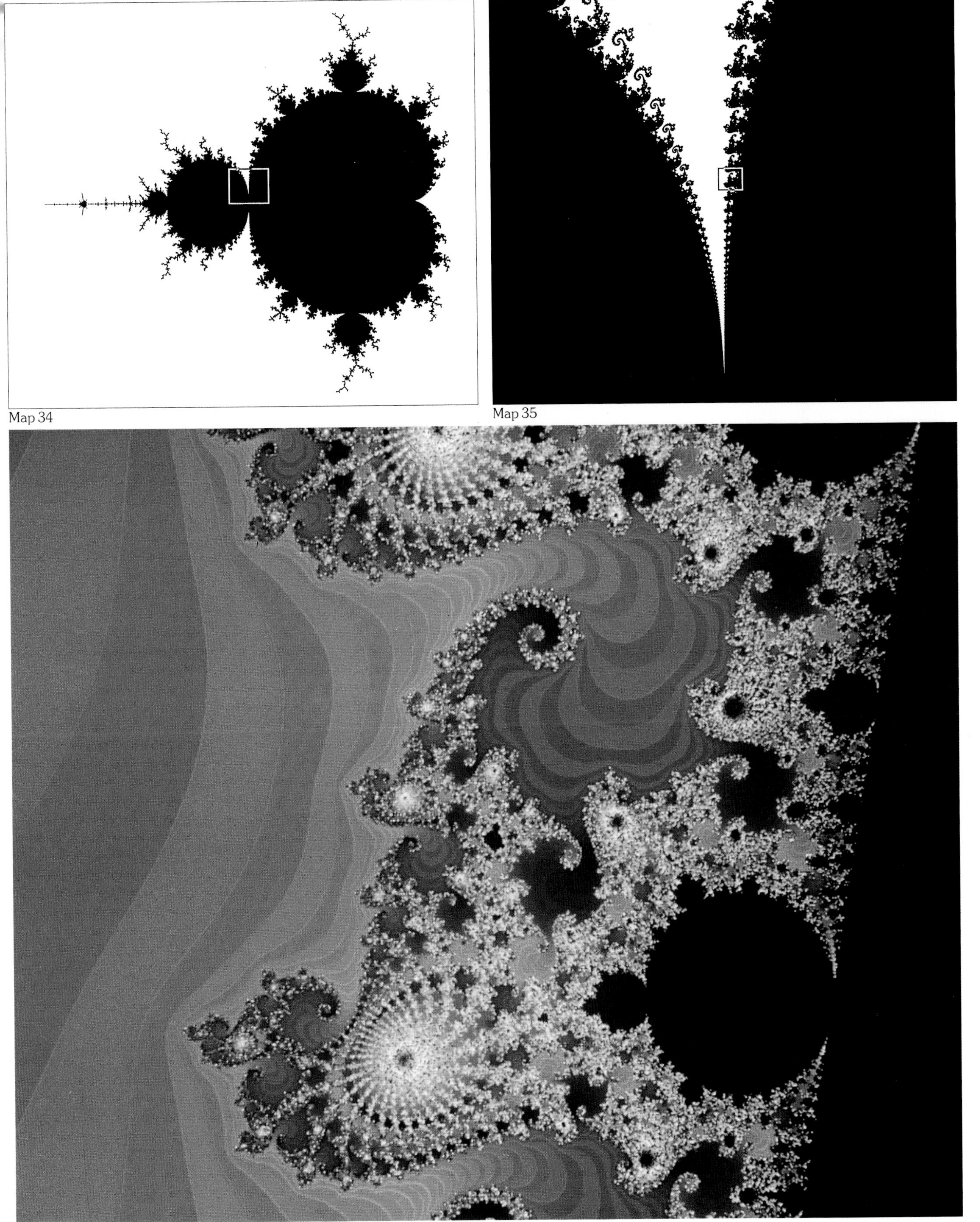

Map 34

Map 35

Map 36

Map 37

Map 38

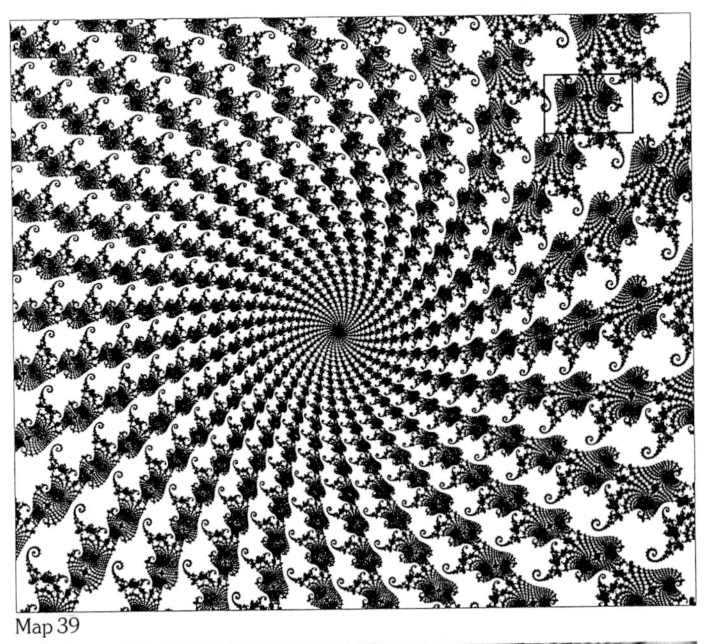

Map 39

Map 40

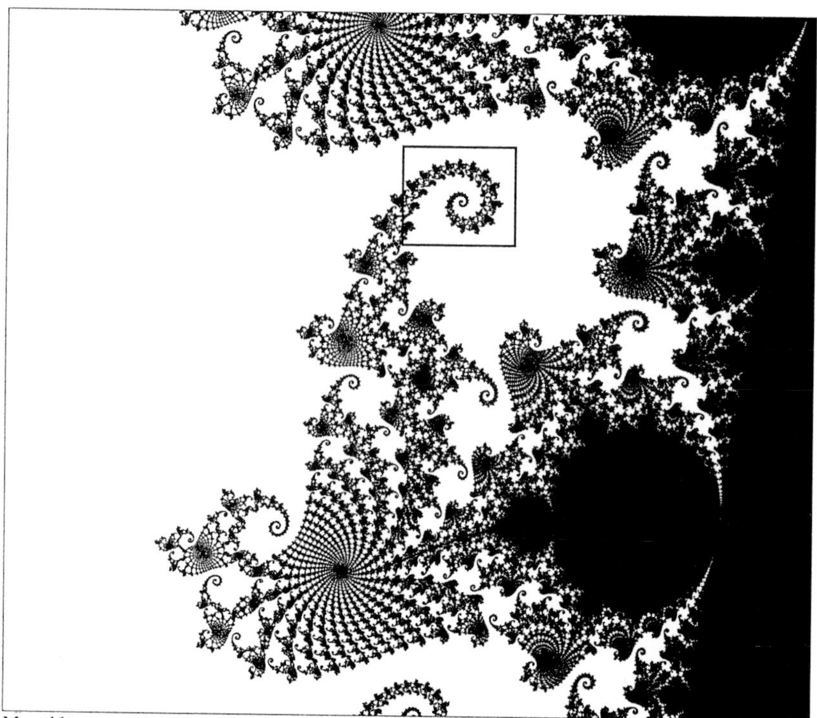

Map 41

Map 42

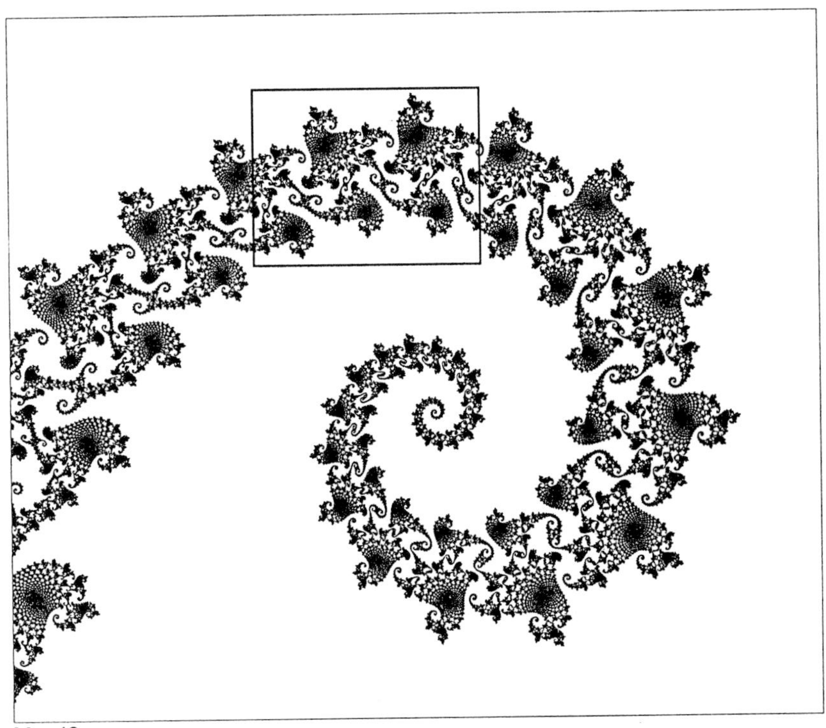

Map 43

Map 44

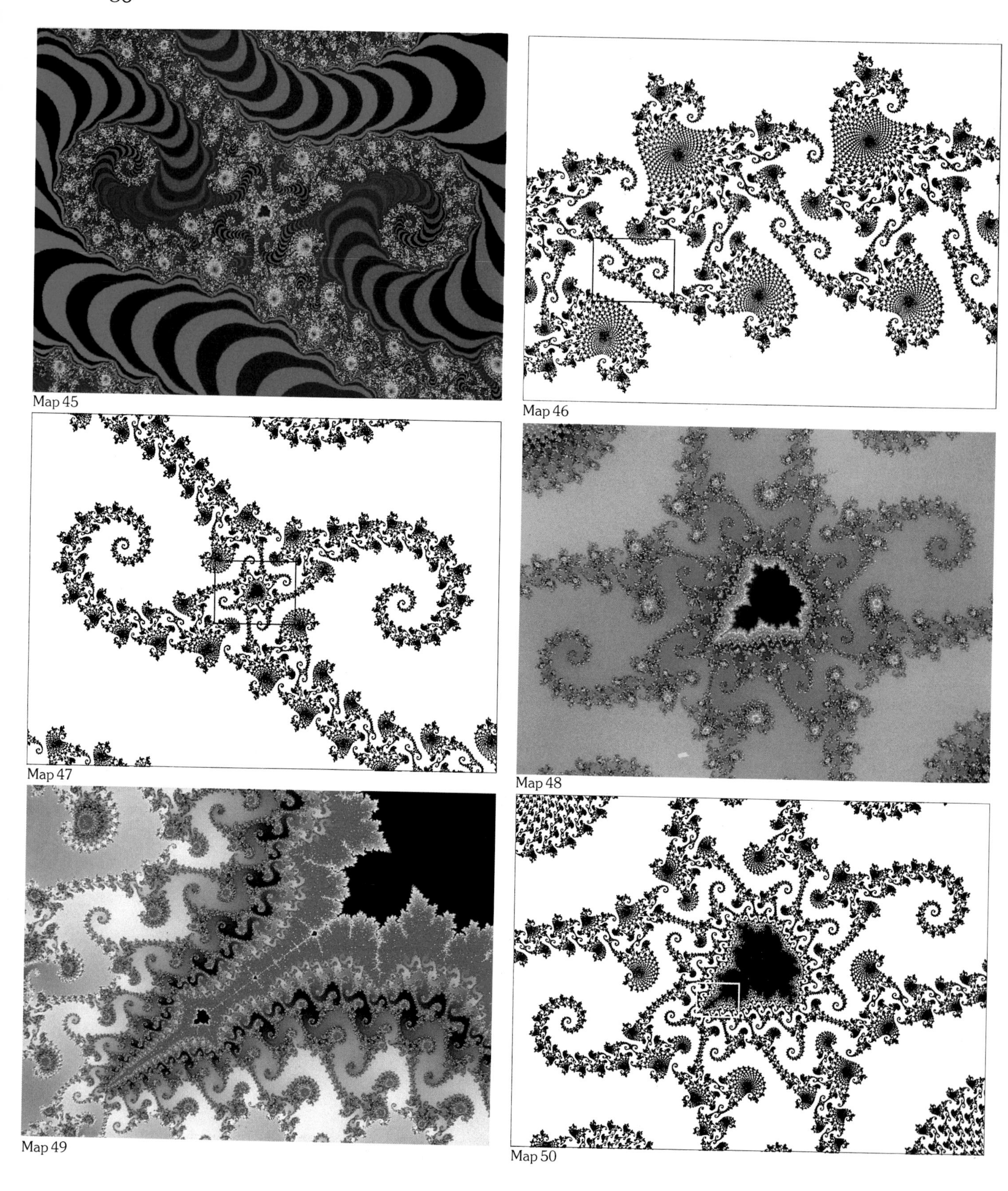

Map 45

Map 46

Map 47

Map 48

Map 49

Map 50

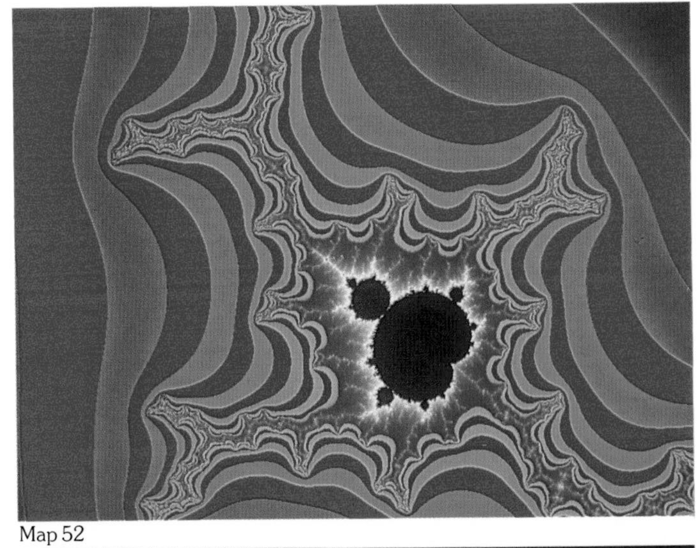
Map 52

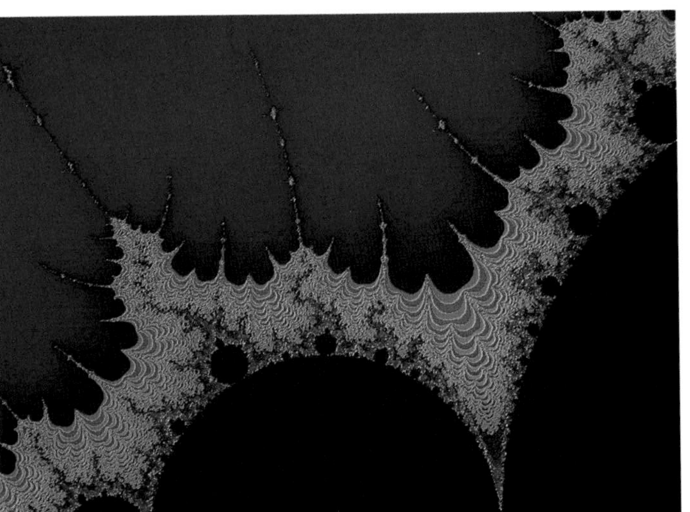
Map 51

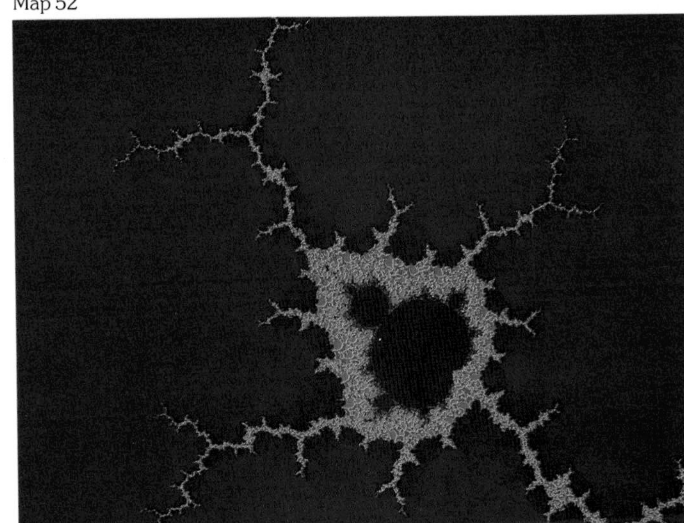
Map 53

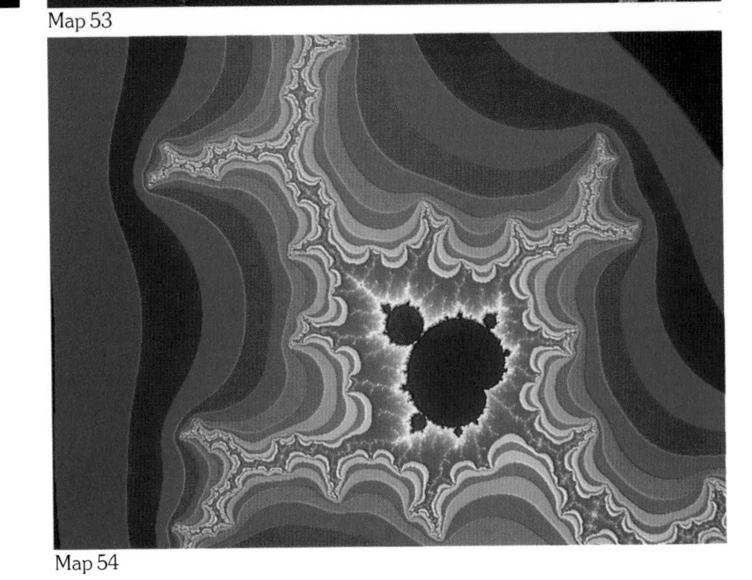
Map 54

Map 55

Map 56

Map 57

Map 58

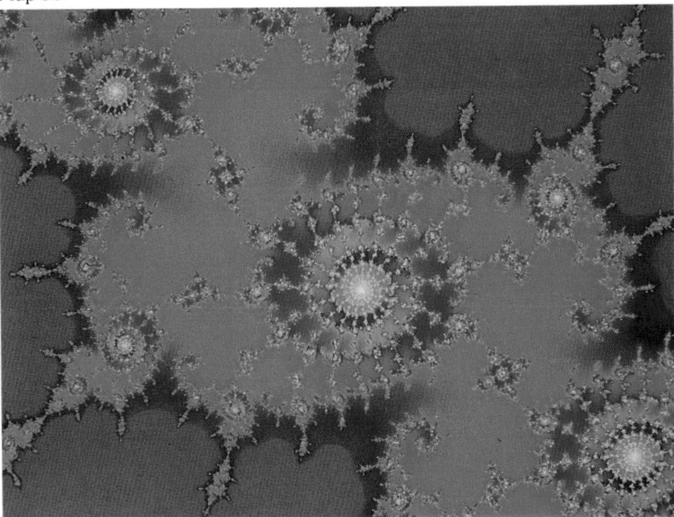
Map 59

Map 60

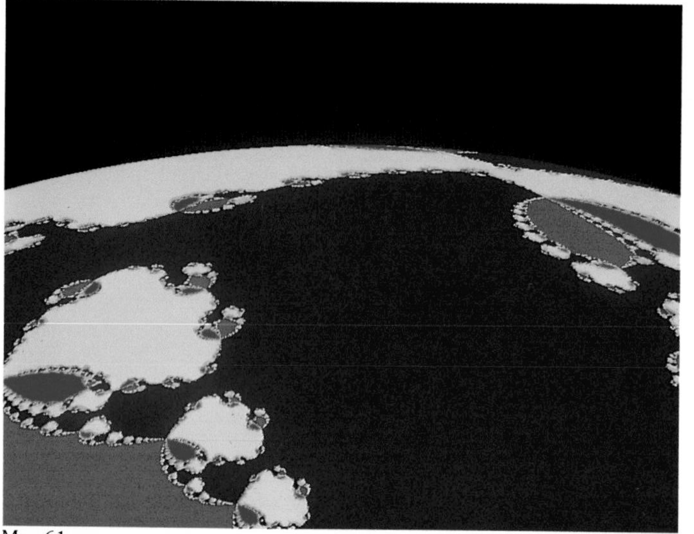

Map 61

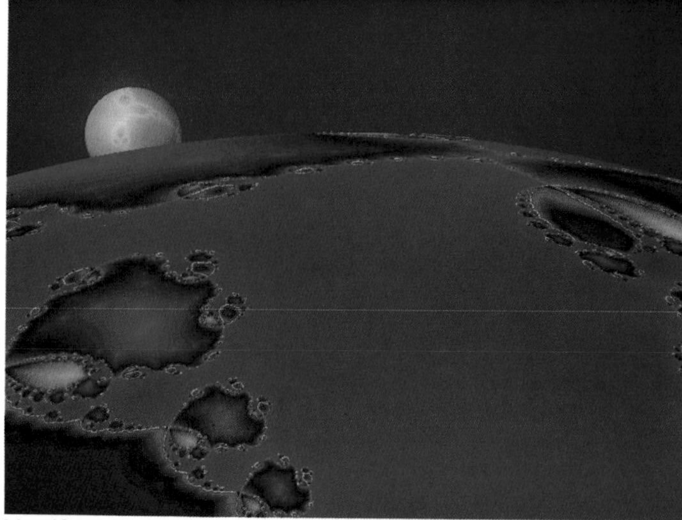

Map 62

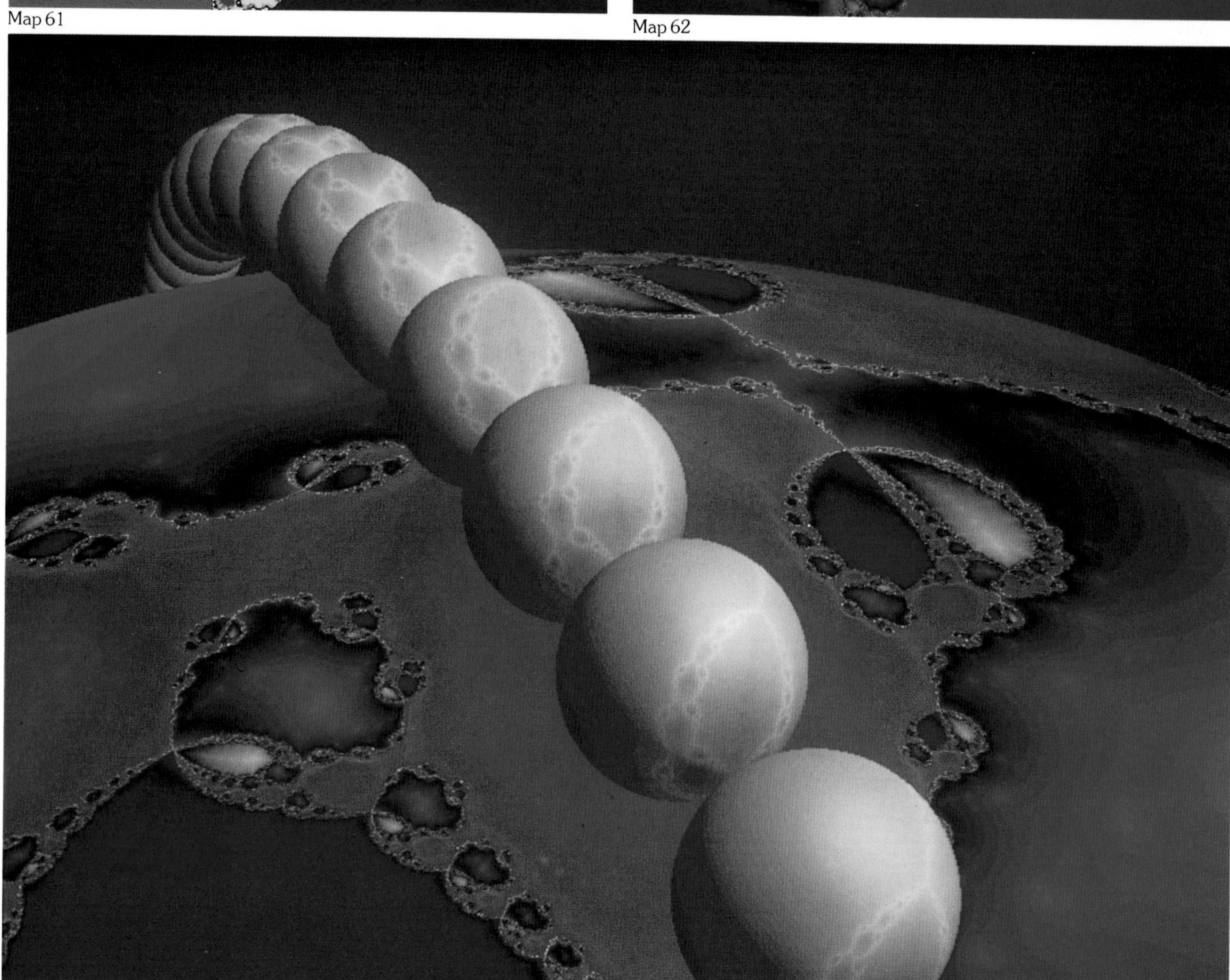

Map 63

Map 64

Map 65

Map 66

Map 67

# 6 Newton's Method for Complex Polynomials: Cayley's Problem

Newton's method and its sophisticated variants are among the most prominent numerical methods for finding solutions of nonlinear equations. The theory of these methods is usually presented in two parts, one with emphasis on the proof of convergence of the method, the other addressing the derivation of the asymptotic speed of convergence. The picture which one thus obtains from the literature is, however, somewhat incomplete. There are additional interesting and deep problems connected with Newton's method one of which is the subject of the following discussion.
The problem we consider now goes back to 1879 and was posed by A. Cayley in a one page paper [Ca] entitled: "The Newton-Fourier imaginary problem". Cayley suggests the extension of what he calls the Newton-Fourier method

$$(6.1)\quad \begin{cases} x_{k+1} = x_k - p(x_k)/p'(x_k) \\ k \quad\;\; = 0, 1, \ldots \end{cases}$$

applied to complex polynomials $p$: "... In connexion herewith, throwing aside the restrictions as to reality, we have what I call the Newton-Fourier Imaginary Problem ..." Furthermore, he suggested to study the problem globally: "... The problem is to determine the regions of the plane, such that $P$ ($=$initial point $x_0$ in (6.1)) being taken at pleasure anywhere within one region we arrive ultimately at the point A ($=$ a root, i.e. $p(A)=0$); anywhere within another region at the point $B$; and so for the several points representing the roots of the equation.
The solution is easy and elegant in the case of a quadratic equation, but the next succeeding case of the cubic equation appears to present considerable difficulty."
From a numerical point of view method (6.1) is always understood to be a local method, i.e. one assumes that $x_0$ (the initial value) is sufficiently close to a root $x^*$ of the equation $p(x^*)=0$. Cayley's problem, instead, is to understand the global basin of attraction for a root $x^*$.

$$(6.2)\quad A(x^*) = \{x \in \overline{\mathbb{C}} : N^k(x) \rightarrow x^* \text{ as } k \rightarrow \infty\},$$

where $N(x) = x - p(x)/p'(x)$ and $N^k = N \circ \ldots \circ N$, $k$ times. Note that

$$(6.3)\quad N'(x) = \frac{p(x)\, p''(x)}{(p'(x))^2},$$

where $' = \dfrac{d}{dx}$. Hence, if $x^*$ is a simple root of $p$ (i.e. $p'(x^*) \neq 0$) we have that $N'(x^*)=0$, i.e. $x^*$ is a superattractive fixed point of $N$. If $x^*$ is a root of multiplicity $k$, i.e. $p(x) = (x-x^*)^k q(x)$ and $q(x^*) \neq 0$, one can easily design a method such that again $x^*$ is a superattractive fixed point:

$$(6.4)\quad N_h(x) = x - \frac{hp(x)}{p'(x)}.$$

Simply choose $h=k$ and observe that $N_h'(x^*)=0$. In general method (6.4) is called a *relaxed* Newton method. Note that (6.4) can also be interpreted as the Euler method with stepsize $h$ for the initial value problem

$$(6.5)\quad \begin{cases} \dot{x}(t) &= -p(x(t))/p'(x(t)) \\ x(0) &= x_0. \end{cases}$$

It is an interesting problem to relate solutions of the differential equation (6.5) to orbits of (6.4) for various choices of $h$.

Cayley's elegant solution for the problem $p(x)=x^2-c$ is worth discussing. (In fact, if $q(x)=ax^2+2bx+d$ is a general polynomial of degree 2, then $q$ is equivalent to $p(x)=x^2-c$ by means of the coordinate change $\Phi(x)=ax+b$, with $c=b^2-b-ad$, i.e. $q(x)=\Phi^{-1}\circ p\circ\Phi(x)$.)

Newton's method for $p$ is the rational map of degree two

$$(6.6)\quad N(x)=(x^2+c)/2x.$$

The claim is that its Julia set $J_N$ is the perpendicular bisector of the segment joining the roots $\pm\sqrt{c}$:

$$(6.7)\quad J_N=\{\alpha i\sqrt{c}:\alpha\in\mathbb{R}\}\ (i=\sqrt{-1}).$$

Thus the basins of the two roots, $A(\sqrt{c})$ and $A(-\sqrt{c})$, are the half planes defined by $J_N$. To see this, one notes that $N$ is conjugate to the simpler map $R(u)=u^2$ by means of the coordinate change $\Psi(x)=(x+\sqrt{c})/(x-\sqrt{c})$, $\Psi^{-1}(u)=\sqrt{c}(u+1)/(u-1)$, i.e. $R(u)=\Psi\circ N\circ\Psi^{-1}(u)$. In other words, whether we study orbits of (6.6) or orbits of $R$ is equivalent. Here is a list of corresponding objects:

(6.8)

| x-plane | $+\sqrt{c}$ | $-\sqrt{c}$ | $\infty$ | $0$ | $J_N$ |
|---|---|---|---|---|---|
| $u$-plane | $\infty$ | $0$ | $1$ | $-1$ | $S^1$ |

$S^1$ is the unit circle $\{u\in\overline{\mathbb{C}}:|u|=1\}$.

The points $u=0$ and $u=\infty$ are attractive fixed points of $R$ with basins $A(0)=\{u\in\overline{\mathbb{C}}:|u|<1\}$, $A(\infty)=\{u\in\overline{\mathbb{C}}:|u|>1\}$. The boundary between these basins, and thus the Julia set of $R$, is $S^1$. It is easy to see that $\Psi^{-1}$ maps $A(0)$ and $A(\infty)$ into the half planes defined by $J_N$. This correspondence shows that $J_N$ is indeed the Julia set of the mapping $N$. As a curiosity we may add that $R(u)=u^2$ is just Newton's method applied to $u\mapsto u/(1-u)$.

### *Dynamics of $u\mapsto u^2$ on $S^1$*

The Julia set of $R(u)=u^2$ illustrates nicely the abstract properties which we have documented in the review on Julia sets. For example, in correspondence with (2.6) and (2.4) we have $\partial A(0)=J_R=\partial A(\infty)$ and $R(J_R)=J_R=R^{-1}(J_R)$. On $S^1=J_R$, $R$ reduces to a 1-dimensional map $r(\alpha)$, $\alpha\in[0,1]$:

$$R(u): u=\exp(2\pi i\alpha)\mapsto u^2=\exp(2\pi i(2\alpha)),$$

$$r(\alpha):\qquad \alpha\mapsto 2\alpha\,(\mathrm{mod}\ 1).$$

In fact $r$ is one of the simplest and best studied mappings which produce *chaotic dynamics*. Properties (2.1) and (2.5) are apparent. Consider

(6.9) $P \;=\{u: u=\exp(2\pi ik/(2^n-1)),\ k,n=1,2,\ldots\}$,

(6.10) $S \;=\{u: u=\exp(2\pi ik/2^n), \qquad k,n=1,2,\ldots\}$.

Both sets are dense in $S^1$. Elements in $P$ are repelling periodic points and $S$ is the inverse orbit of 1, $Or^-(1)$. To describe the dynamics of $r$ completely is a delicate problem. A typical difficulty arises when we try to determine those $\alpha$ for which the forward orbit $Or^+(\alpha)$ is dense, i.e. comes arbitrarily close to any point in $[0,1]$. At first one would guess that any irrational $\alpha$ has this property. But that is not true. To obtain a counterexample consider the following representation of the action of $r$. Let $\alpha=(a_1, a_2, a_3, \ldots)$ be the binary expansion of $\alpha\in[0,1]$, i.e.

$$\alpha = \sum_{k=1}^{\infty} a_k 2^{-k},\quad a_k\in\{0,1\}.$$

Then $r(\alpha)=(a_2, a_3, a_4, \ldots)$, i.e. $r$ acts as a shift. Note that due to the finiteness of machine numbers an iteration of the shift $r$ in a computer simulation will *always* approach 0, which is quite a different behavior from the true dynamics of $r$. This shows that computer experiments can be dramatically misleading. Now let

$$(6.11)\qquad \alpha^* = \sum_{n=1}^{\infty} 2^{-n!}$$

Then $Or^+(\alpha^*)$ is *not* dense in $[0,1]$. Indeed, the only cluster points are 0 and $2^{-k}$, $k=1, 2, 3, \ldots$. ($\alpha^*$ is an example of a Liouville number.) So are there dense orbits at all? The reader may verify that the following choice meets the goal:

$$(6.12)\qquad \bar\alpha=(\,\underbrace{.1}_{1}\,,\ \underbrace{1,0}_{2}\,,\ \underbrace{1,1}_{3}\,,\ \underbrace{1,0,0}_{4}\,,\ \ldots,\ \text{binary exp. of } n\,,\ldots).$$

Thus, the dynamics of $R$ on $J_R$ is rather mysterious, and that is in general the case with Julia sets.

### *Newton's Method for $x^3-1$*

In the attempt to extrapolate from our experience with Newton's method for $p(x)=x^2-1$ one might expect that the Julia set for

$$(6.13)\qquad N(x)=x-\frac{x^3-1}{3x^2}=\frac{2x^3+1}{3x^2}$$

is the set $S$ shown in Fig. 39.
Indeed, one observes that $N(x)=D\circ N\circ D^{-1}(x)$, where $D$ is the rotation by 120°. Furthermore $N(\infty)=\infty$ is a repelling fixed point and $N(0)=\infty$, $N(x_k)=0$ for $x_k=-\exp(2\pi ik/3)\sqrt[3]{2}$. Thus all these points $\infty, 0, x_0, x_1, x_2$ are in the Julia set $J_N$ of $N$.

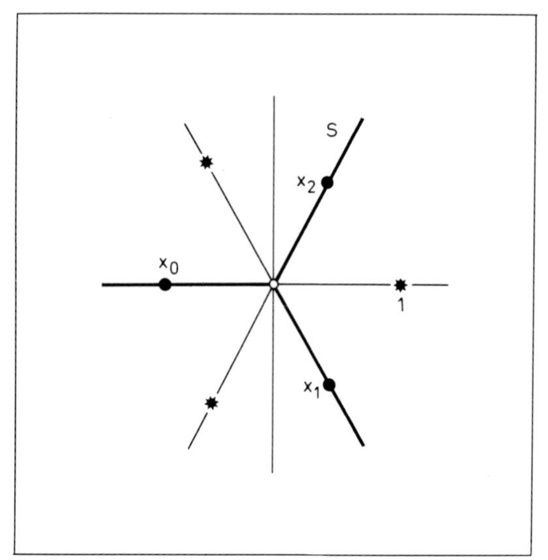


Fig. 39. Fixed points of N and preimages of 0

Fig. 40. Separatrix for the Newton flow

Furthermore, one can easily show that the phase portrait of (6.5) is as indicated in Fig. 40. The set $S$ there serves as a separatrix for the three stable regions determined by the third roots of unity. Indeed (6.5) is equivalent to the (real!) initial value problem

$$\begin{cases} \alpha\dot{x}(t) = -p(x(t))\cdot\overline{p'(x(t))}, & \alpha = |p'(x(t))|^2 \\ x(0) = x_0, \end{cases}$$

where $\overline{a+ib} = a - ib$. Thus, the point 0 is distinguished in that it is a (and the only) three-corner-point with regard to the stable regions of (6.5).

The surprise is that we are *fundamentally wrong* in guessing that S is the Julia set for (6.13). This was known already to *Julia* and *Fatou*. Had we recalled property (2.6) from the review on Julia sets, we could not have been misled:

(6.14) $\quad J_N = \partial A(\exp(2\pi ik/3)),\ \ k = 0, 1, 2;$

i.e. each point in $J_N$ must be a three-corner-point with respect to the basins of attraction of the third roots of unity. Let us give a short argument for (6.14) using property (2.5) i.e.

(6.15) $\quad \{x : N^k(x) = 0 \text{ for some } k\}$ is dense in $J_N$.

Let $x \in J_N$ and $U$ be an arbitrarily small neighborhood of $x$. Due to (6.15) we find $\bar{x} \in U \cap J_N$ such that $N^k(\bar{x}) = 0$ for some $k$. Note that $(N^k)'(\bar{x}) \neq 0$ (indeed, the only points $x$ with $N'(x) = 0$ are the third roots of unity), and therefore $N^k$ is invertible in a small neighborhood of $\bar{x}$ (see Fig. 41). Thus we find a neighborhood $V$ of $\bar{x}$ and an $\varepsilon$-ball $B_\varepsilon(0)$, such that $N^k$ is $1-1$ and onto from $V \to B_\varepsilon(0)$. Now let $I_\varepsilon$ be the interval $I_\varepsilon = (0, \varepsilon) \subset \mathbb{R}$. Then it is easy to see (check the graph of $p(x) = x^3 - 1$, $x \in \mathbb{R}$, and the geometric interpretation of Newton's method) that $I_\varepsilon \subset A(+1)$. Consequently, due to the symmetry in $N$,

$$D(I_\varepsilon) \subset A((-1 + i\sqrt{3})/2) \text{ and } D^2(I_\varepsilon) \subset A((-1 - i\sqrt{3})/2),$$

Fig. 41. $N^k$ transplants the three-corner-point 0 (R, G, B = Red, Green, Blue) along the Julia set

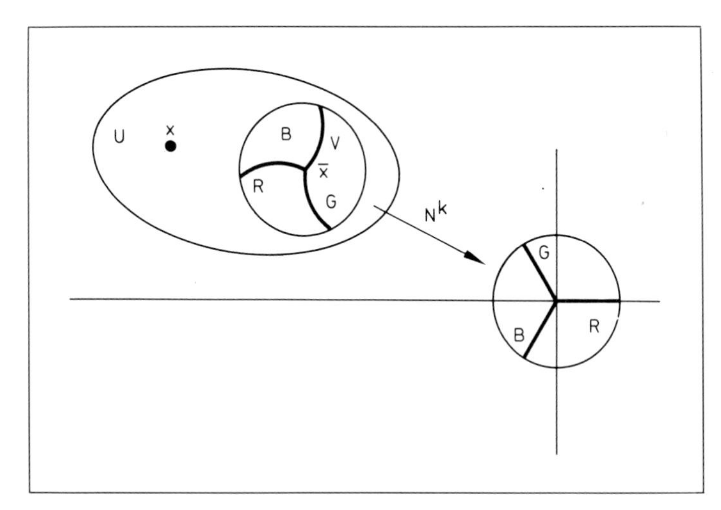


Fig. 42. Basin of attraction $A(1)$ in white ▽

which proves that the three basins of attraction have points in $U$. Now $U$ was arbitrary and, hence, (6.14) is established. Recall that for the continuous system (6.5) the origin 0 was the only three-corner-point. Our argument above thus means that discretizing (6.5) by Euler's method spreads out this distinguished point infinitely many times and creates thus the Julia set $J_N$ as a separatrix for the three basins of attraction. Figure 42 shows a portion of $A(1)$ in white and Fig. 43 the Julia set in a similar window; the three successive close-ups reveal a remarkable self-similarity of crab-like structures.

Figure 44 shows a binary decomposition of $A(1)$ obtained according to (2.37) for $\alpha_0 = 0$, and replacing $\frac{1}{2\pi} \arg R^k(x)$ by $\frac{1}{2\pi} \arg(N^k(x) - 1)$ ($N$ as in (6.13)).

Actually similar experiments led us to a rediscovery of Boettcher's result (2.34), i.e. in a neighborhood of 1 Newton's method $N$ for $p(x) = x^3 - 1$ is equivalent to the dynamics of $u \mapsto u^2$. In fact, since $N$ has no other critical point in $A^*(1)$ (the immediate basin of attraction of the root 1) than 1 itself, one can show that this conjugation extends to all of $A^*(1)$. Therefore, if $\Phi$ is

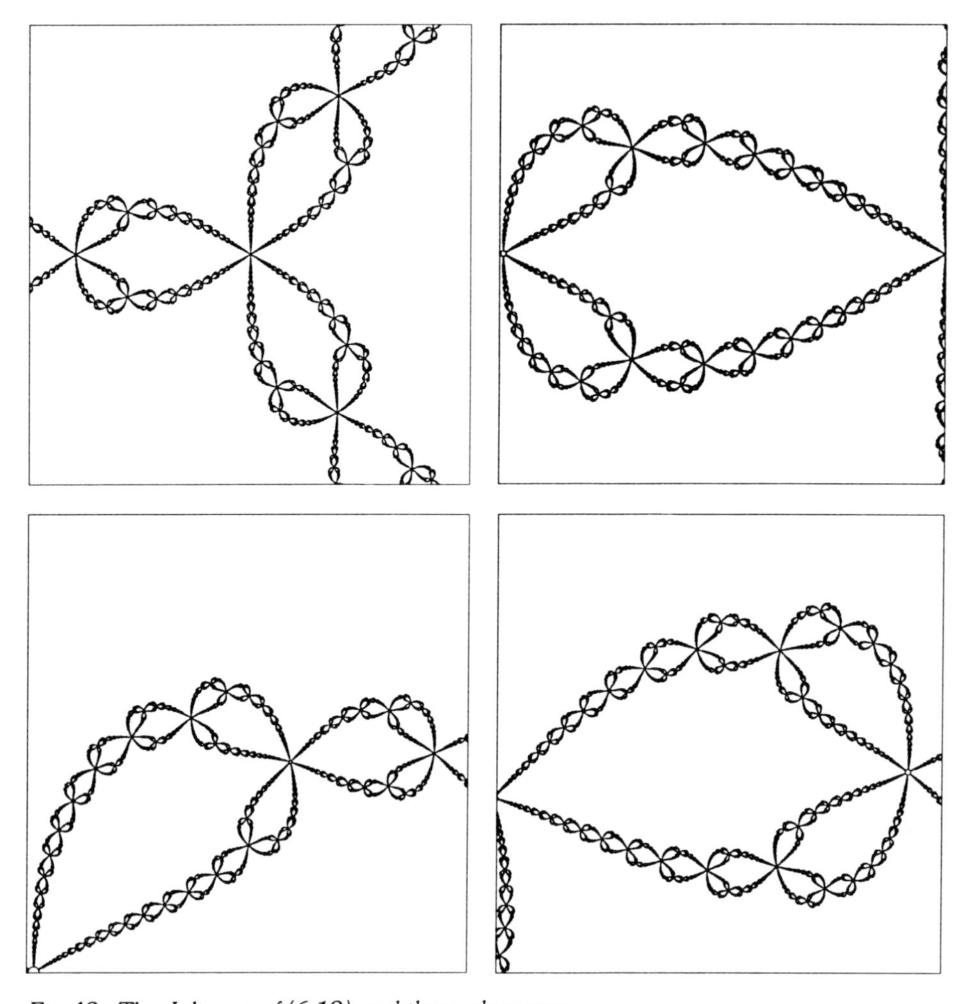

*Fig. 43. The Julia set of (6.13) and three close ups*

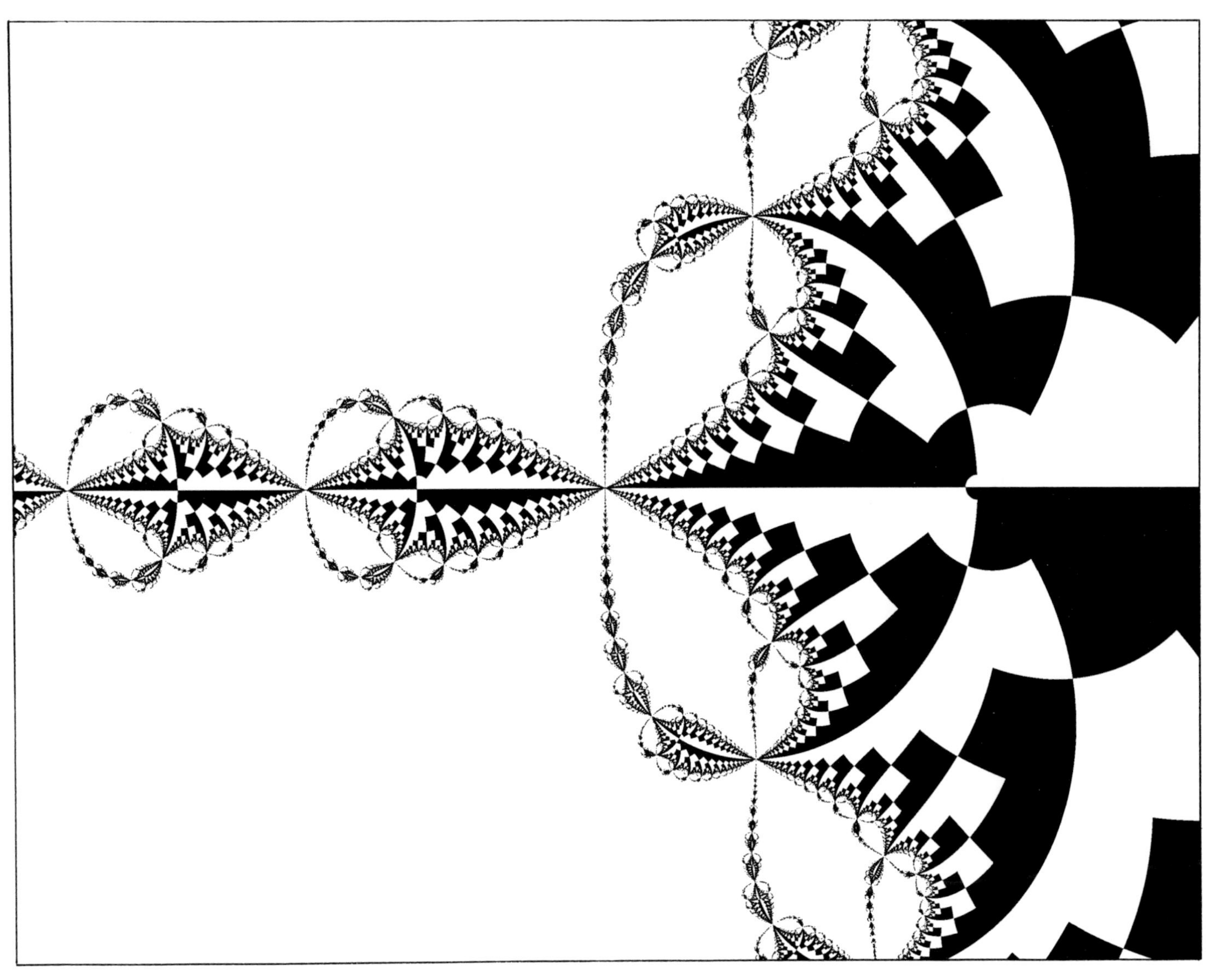

*Fig. 44. Binary decomposition of $A(1)$ for Newton's method (6.13) (see also Map 66)*

the conjugation in (2.34), i.e. $\Phi(0)=1$, $\Phi'(0)=1$ and $\Phi^{-1}\circ N\circ \Phi(u)=u^2$, then $\Phi$ induces a system of internal angles to $A^*(1)$. In this way Fig. 44 indicates the binary addresses for each $x\in\partial A^*(1)$ by following internal angles. Map 66 shows a binary decomposition for all basins of $N$.
As we outlined above, Newton's method for second degree polynomials is completely understood, and there the general case is equivalent to the special case of $p(x)=x^2-1$. Now that we have mastered the case $p(x)=x^3-1$ also, we might guess that this case is *typical* for *all* polynomials of degree three. That this is far from being true has recently been reported in a study which also made a surprising connection to the Mandelbrot set: We discuss this finding next.

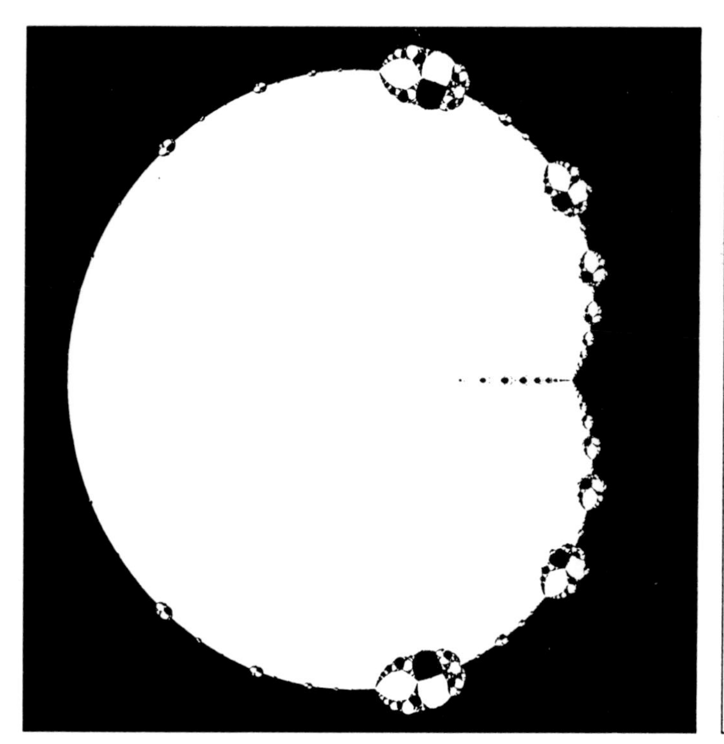

*Fig. 45. Study of the critical point* 0 *in the complex $\lambda$-plane; if $\lambda$ is from the black set, then* $0 \in A(1)$; right: *a close up from* left

## *Newton's Method for Arbitrary Polynomials of Degree 3*

A remarkable result of B. Barna [Ba] states that for real polynomials with only real roots Newton's method will converge to a zero, for almost every starting value taken from $\mathbb{R}$. In our previous section we observed that Newton's method (6.13) for $p(x)=x^3-1$ has a similar property, i.e. except for initial values from a set of measure zero (the Julia set), we have convergence to a root of unity. In general, however, this may not be true, i.e. there may be complex polynomials for which Newton's method allows for sets of initial values of full measure, such that one does *not* find convergence to a root. Such examples have been obtained in a computer assisted study by J. Curry, L. Garnett and D. Sullivan [CGS].
The first step in the argument is that Newton's method for any cubic polynomial is equivalent, by an affine change of variables, to Newton's method for

(6.16) $p_\lambda(x)=x^3+(\lambda-1)x-\lambda$

with $\lambda$ suitably chosen. Maps 77 and 78, which by the way are based on the same data, and thus demonstrate the effect of different color maps, show parts of three basins of attraction for the choice $\lambda=0.5$. Newton's method $N_\lambda$ for (6.16) is a rational function of degree three with 4 critical points: the 3 roots of $p_\lambda$ and the distinguished point 0. We now use Sullivan's classifica-

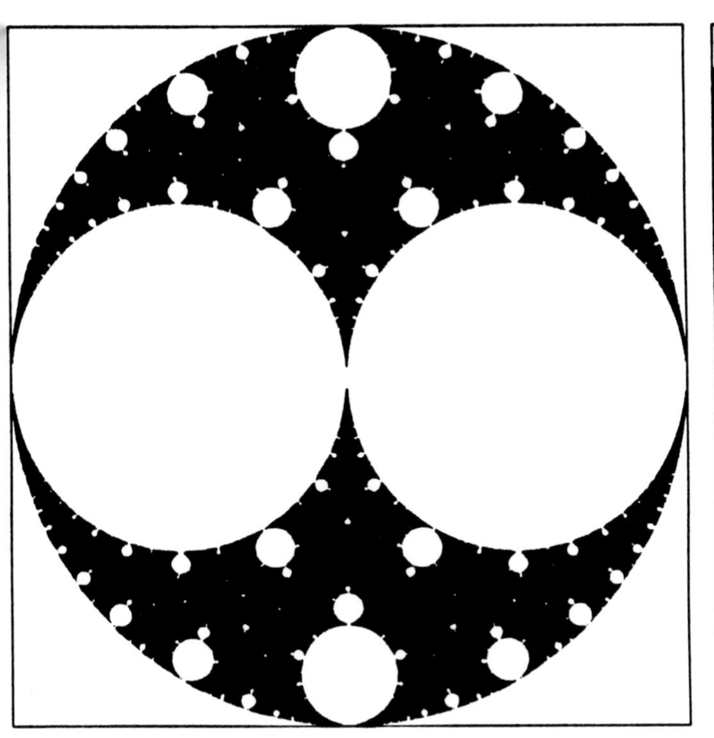

*Fig. 46. Study of the critical points $\pm 1$ of $u \mapsto \lambda(u+1/u)$ in the complex $\lambda$-plane; for $\lambda$ from the black set, $\pm 1$ are not attracted by an attractive periodic orbit;* right: *a close up from* left

tion (3.5) to find parameter values $\lambda$ for which the critical point 0 is *not* in the basin of attraction of any of the 3 roots of $p_\lambda$. In other words, let

(6.17) $M=\{\lambda \in \overline{\mathbb{C}}: N_\lambda^k(0) \nrightarrow \text{root of } p_\lambda, \text{ as } k \to \infty\}$.

Figure 45 shows computergraphical studies for the critical point 0 in which the Mandelbrot-like set $M$ contains parameters $\lambda$ for which $N_\lambda$ has attractive periodic points, and thus manifests counterexamples to Barna's result in the complex case.

## The Relaxed Newton Method

We noted already that the relaxed Newton method (6.4) can be interpreted as a numerical discretization of the differential equation (6.5). In this final section we try to understand (6.4) for varying *stepsize* $h$ in the special case of polynomials of degree two. We know already that it suffices to understand $p(x)=x^2-c$, which leads to the relaxed Newton method

(6.18) $N_h(x)=x-h\dfrac{x^2-c}{2x}$.

For $h=0$ and $h=2$ one obtains the trivial functions $N_0(x)=x$ and $N_2(x)=\dfrac{c}{x}$.

So let us assume $h \in \mathbb{C} \setminus \{0,2\}$. Introducing new coordinates

$$x = -i\sqrt{c}\sqrt{\frac{h}{h-2}}\, u,$$

we obtain $N_\lambda(u) = \lambda(u + \frac{1}{u})$ for $\lambda = \left(1 - \frac{h}{2}\right)$. B.Mandelbrot investigated this family in his book [Ma]. Our Fig.46 is a computergraphical study of the behavior of the critical points $u = \pm 1$ of $N_\lambda$: The set of parameters $\lambda$ for which the critical points of $N_\lambda$ do not tend to an attractive periodic orbit is shown in black. We have *experimental evidence* that an attractive periodic orbit exists in white regions.

# 7 Newton's Method for Real Equations

Much of the complexity which we have seen in Newton's method for complex polynomials is known to be closely linked to the underlying complex analytic structure. Thus, it appears to be an interesting question to ask what the situation is like for systems of real equations. Note, however, that a complex analytic map $\mathbb{C} \ni x \mapsto f(x) \in \mathbb{C}$ can be regarded as a function of two real variables in a canonical way, viz. $f(x) = (f_1(x_1, x_2), f_2(x_1, x_2))$ such that the Cauchy-Riemann differential equations are satisfied:

$$(7.1) \quad \frac{\partial f_1}{\partial x_1} = \frac{\partial f_2}{\partial x_2}, \; \frac{\partial f_1}{\partial x_2} = -\frac{\partial f_2}{\partial x_1}$$

Thus, if we want to address the above question, we should consider mappings that do not satisfy (7.1). More generally, let $H: \mathbb{R}^n \to \mathbb{R}^n$, where $\mathbb{R}^n$ is the n-dimensional Euclidean space, be a smooth mapping. Then a damped version of Newton's method for solving $H(x) = 0$ is

$$(7.2) \begin{cases} x_{k+1} = N(x_k) = x_k - h[DH(x_k)]^{-1} H(x_k), \\ k = 0, 1, 2, \ldots, \; x_k \in \mathbb{R}^n. \end{cases}$$

for $h \in \mathbb{R}$.
Note that (7.2) can be interpreted as Euler's method with stepsize $h$ for the initial value problem

$$(7.3) \begin{cases} \dot{x}(t) = -[DH(x(t))]^{-1} H(x(t)), \\ x(0) = x_0. \end{cases}$$

As in the complex case (cf. the Special Section 6.) we are particularly interested in the relation of (7.2) and (7.3), i.e. we want to understand (7.2) as $h$ is varied. The article [PPS] addresses this question at length. Here we collect a few results which allow us to interpret some of our computergraphical experiments.

Let us begin by posing some immediate questions:

(7.4) What is the candidate of a Julia-like set for the dynamical system (7.2)?

(7.5) Which properties of such Julia-like sets are shared with Julia sets of rational maps on $\overline{\mathbb{C}}$, and which are not? In particular, is property (2.6) satisfied, and what is the dynamics on the Julia like set?

(7.6) Are Julia-like sets typically fractal sets?

Turning to question (7.4), we recall that for a polynomial $p$ the Julia set of Newton's method is obtained from the iterated preimages (inverse orbits) of zeros of $p'(x)$. Thus, we are led to focus here on the set

$$(7.7) \quad S = \{x \in \mathbb{R}^n : det(DH(x)) = 0\}.$$

Typically, $S$ is a collection of $(n-1)$-dimensional manifolds in $\mathbb{R}^n$, and a reasonable candidate for a Julia-like set appears to be

(7.8) $\text{closure}\ \{x \in \mathbb{R}^n : N^k(x) \in S \text{ for some } k \geq 0\}$,

(closure of X=X together with all cluster points of X), where $N$ is Newton's map associated with $H$ as given in (7.2). Another equally reasonable candidate is

(7.9) $\text{closure}\ \{x \in \mathbb{R}^n : \|N^k(x)\| \to \infty \text{ as } k \to \infty\}$.

The generating sets of (7.8) and (7.9) are clearly different, but together with all their accumulation points they may become identical. In fact, this is one of our conjectures. To explain our experiments it remains to describe the choice of $H$:

(7.10) $H(x) = Mx - \mu F(x)$,

where $\mu$ is a real parameter, $x = (x_1, \ldots, x_n)^T$, $F(x) = (f(x_1), \ldots, f(x_n))^T$, $f: \mathbb{R} \to \mathbb{R}$, and $M$ is the matrix

$$\begin{pmatrix} 2 & -1 & & 0 \\ -1 & \ddots & \ddots & \\ & \ddots & \ddots & -1 \\ 0 & & -1 & 2 \end{pmatrix}$$

Without going into any details, we merely note that (7.10) is obtained as a finite difference approximation for the two-point boundary value problem

$$(7.11) \begin{cases} u''(t) + \lambda f(u(t)) = 0, \\ u(0) = 0 = u(1), \end{cases}$$

for $n$ internal meshpoints in (0,1) with meshwidth $1/(n+1)$, and $\mu = \lambda/(n+1)^2$. The parameter $\lambda$ serves as a *bifurcation parameter* for the boundary value problem if $f$ is a nonlinearity of a suitable kind [PPS]. Our choices for $f$ are:

(7.12) $f(s) = s - s^2$ and $f(s) = s - s^3$.

Choosing $H$ as in (7.10), the mapping $N$ in (7.2) depends on two external parameters $h$ and $\mu$. Whenever necessary we will stress this by writing $N_{h,\mu}$. It is not hard to show that any solution of $H(x) = 0$ is a stable center for (7.3). Figures 47 and 48 provide the phase portraits for the continuous flow (7.3), using two characteristic choices of $\mu$, with $f(s) = s - s^2$ and $n = 2$. If $1 < \mu < 3$, then $H(x) = 0$ has two solutions $x = a_1$ and $x = a_2$, one of which is $a_2 \equiv 0$. If $\mu$ passes beyond 3, then the solution $a_2 \equiv 0$ bifurcates into 3 solutions, i.e. for $\mu > 3$ we have 4 solutions. The singular set $S$ in (7.7) depends on $\mu$, but is always given by two hyperbolas, the lower one being denoted by $S^-$ and the upper by $S^+$. If $1 < \mu < 3$, then $S^-$ *absorbs* orbits and $S^+$ *repels* orbits, while for $\mu > 3$, $S^+$ still *repels* but now $S^-$ *absorbs* in some parts and *repels* in others. The transition is understood as a bifurcation. The straight line $G$ plays a particular role (see Fig. 47). If $1 < \mu < 3$, then all orbits of (7.3) with initial

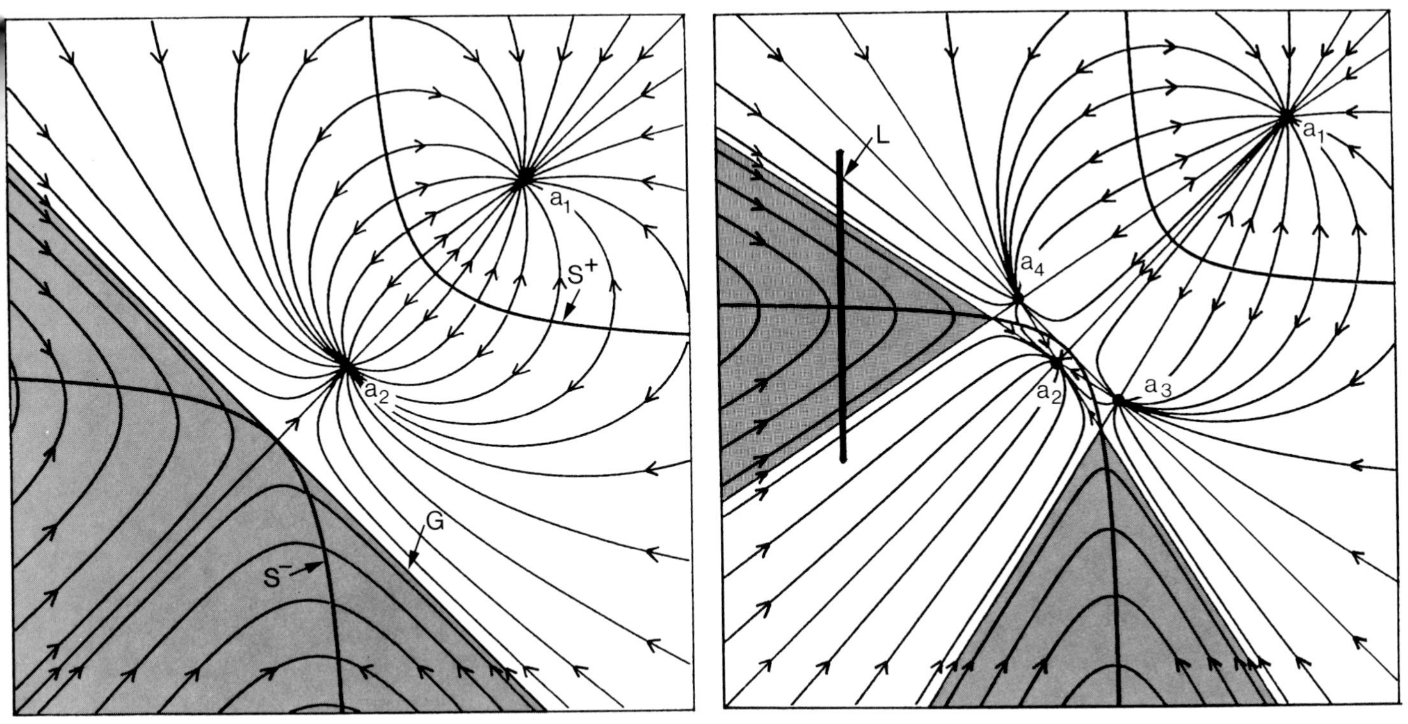


*Fig. 47. Flow of (7.3) for $f(s) = s - s^2$ and $1 < \mu < 3$*

*Fig. 48. Flow of (7.3) for $f(s) = s - s^2$ and $\mu > 3$*

points $x_0$ below $G$ will in finite time arrive in $S^-$, and $G$ itself is invariant, i.e. for $x_0 \in G$ one has that $x(t) \in G$. Similarily, if $\mu > 3$ we find conical regions bounded by straight lines such that any orbit of (7.3) with initial point in that region will arive in $S^-$. Thus, one is led to expect that these regions, which are shaded in Figs. 47 and 48, should not belong to basins of attraction of roots $H(x) = 0$ when (7.3) is discretized by Euler's method, (7.2). It is quite a surprise that this intuition is simply wrong for any $h > 0$.

Let us first discuss the case $1 < \mu < 3$. As is apparent from Fig. 47, solution $a_1$ of $H(x) = 0$ attracts any orbit of (7.3) with initial point above $S^+$, while solution $a_2$ of $H(x) = 0$ attracts any orbit of (7.3) with initial point below $S^+$ and above $G$. For the discrete system (7.2) things change dramatically. While $a_1$ and $a_2$ are of course attractive fixed points for $N$ (provided $0 < h < 2$), and the basin of $a_1$ is exactly the region above $S^+$ (as suggested by the phase portrait of (7.3)), the basin of $a_2$ extends into the half-plane below $G$. In fact $A(a_2)$ now has infinitely many components bounded by the set (7.8). Figure 49 shows the Julia-like set (7.8) for various choices of $h$, and this explains Map 67, where $h = 0.1$. The region below $G$ is almost densely covered by the Cantor set of curves (7.8), but thin filaments of components of $A(a_2)$ are interspersed in that region, resulting in a *noisy structure* below $G$. The layer structures that we see above $G$ in Map 67 are *level sets of equal dynamic distance* from the attractors $a_1$ and $a_2$ respectively.

It is noteworthy that also $N$ leaves $G$ invariant, and when restricted to $G$, it is a

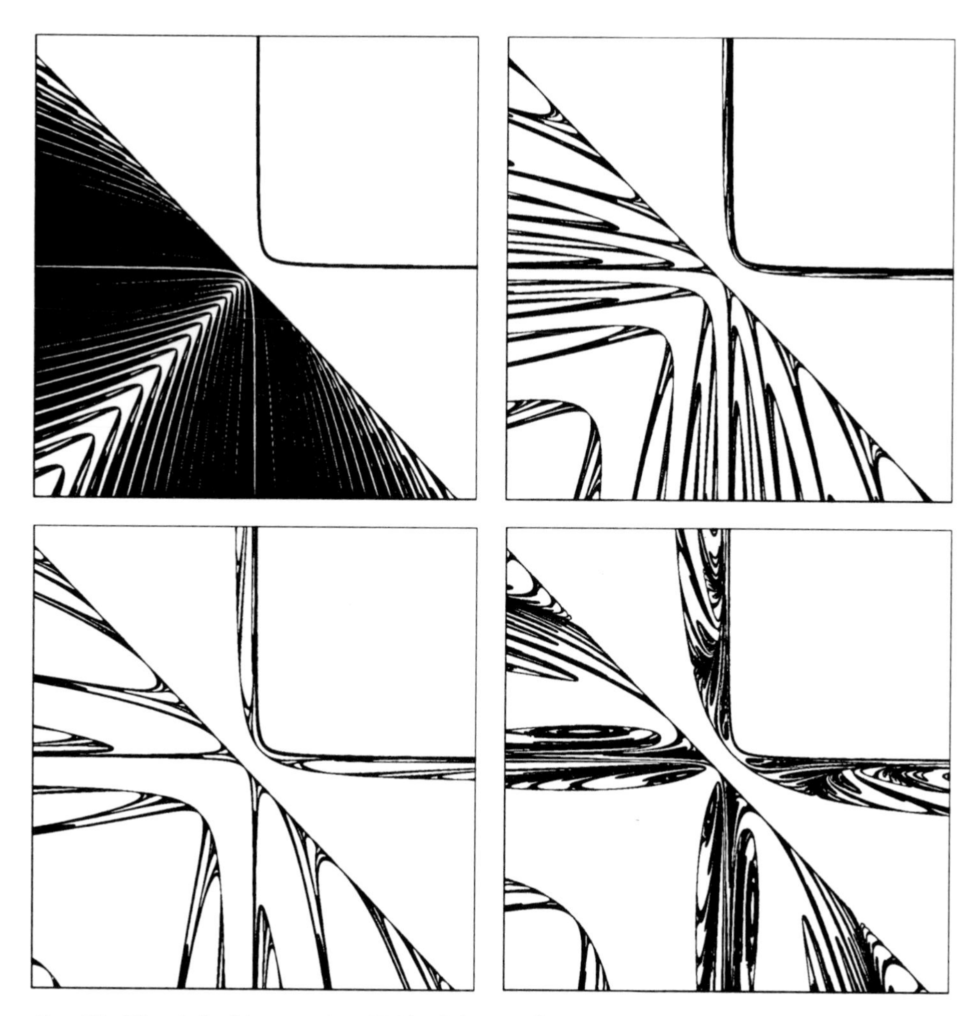

*Fig. 49. The Julia-like set for (7.2), $f(s)=s-s^2$, $\mu=2.1$.* Above left: $h=0.3$; above right: $h=1.0$; below left: $h=1.4$; below right: $h=1.7$ *(see also Maps 17 and 67)*

one-dimensional mapping equivalent to a relaxed Newton method for the polynomial

(7.13) $p(s)=\mu s^2-(\mu+1)(\mu-3)/4\mu.$

Thus on $G$ the dynamics of $N$ is completely described by (6.18) and the Mandelbrot-like sets in Fig. 46. In particular, the dynamics is chaotic on $G$ for any $0<h<2$, and $1<\mu<3$. Moreover, if $h=1$ and $1<\mu<3$ then the dynamics of $N$ restricted to $G$ is equivalent to

$$\alpha \mapsto 2\alpha (mod\ 1)\ \alpha\in[0,1]$$

or to $u \mapsto u^2$ on the unit circle. It is quite remarkable how this simplest of all nonlinear maps pervades all discussions of complex dynamics.
If $\mu>3$, two more attractive fixed points $a_3$ and $a_4$, together with their basins of attraction enter the scene to compete for the plane with the already existing solutions $a_1$ and $a_2$. (Except for $a_2\equiv 0$ all other solutions $a_1$, $a_3$ and $a_4$ de-

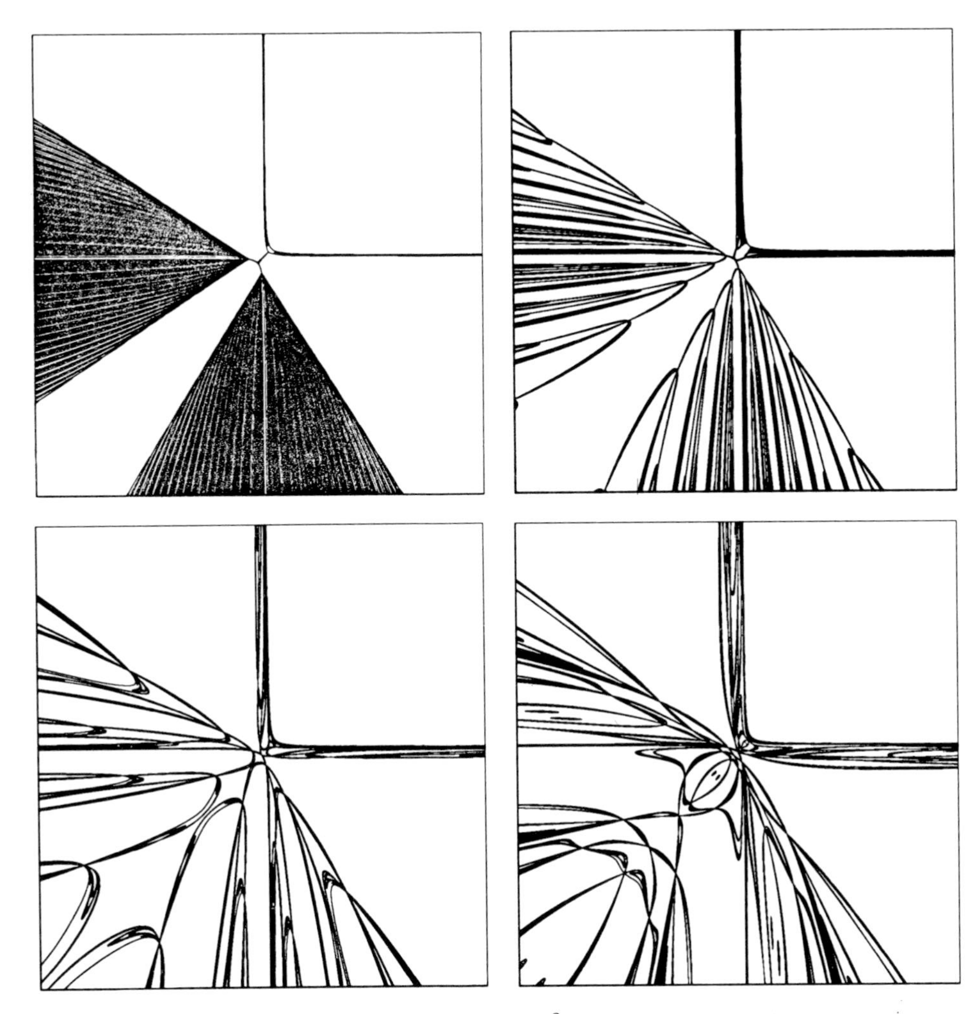

*Fig. 50. The Julia-like set for (7.2), $f(s)=s-s^2$, $\mu=3.2$.* Above left: $h=0.3$; above right: $h=1.0$; below left: $h=1.4$; below right: $h=1.7$ *(see also Maps 19, 21, 23, 68, 71–73, 87)*

pend on $\mu$, and as $\mu$ changes they establish continuous branches.) Having now four basins one might expect, from previous experience with Newton's method for complex polynomials, that there is a Julia-like set which consists of four-corner-points. Maps 19, 21 and 23 show the result of experiments for $h=0.2$, $h=1.8$ and $h=1.9$. Red is used for the basin of $a_1$, yellow for $a_2$, green for $a_3$ and blue for $a_4$ (compare with Fig. 48 for the location of the four roots). Figure 50 shows the Julia-like sets (7.8) for the respective choices of $h$. Map 68 is derived from the same data base as Map 19 except that we have chosen a different color map which stresses level sets in the basins.
Two experimental findings are contrary to our expectations. Though there are four basins, only three of them (yellow, green and blue) seem to be getting entangled and creating boundaries made up of three-corner-points. The fourth (red) is bounded by $S^+$ and is identical with the respective basin in the continuous case (7.3), see Fig. 48. The other irritating finding is that even for

small $h$ the conical sections (shaded in Fig. 48) are now providing initial points for which (7.2) converges to a root of H, and that is quite different from the continuous case (7.3) in which these sections are *absorbed* by $S^-$ in finite time.

To investigate what precisely happens in these conical sections we performed another experiment which resulted in Maps 69, 70 and 88. For that experiment we chose a vertical line segment $L$ which intersects one of the conical sections as shown in Fig. 48.

(7.14) $L = \{(x_1, x_2) \in \mathbb{R}^2 : x_1 = -5, \ -5 \leqslant x_2 \leqslant 5\}.$

Then for each $x \in L$ and $0.002 \leq h \leq 1.998$ we computed the forward orbit $Or^+(x)$ according to (7.2) and used colors to exhibit the result:

$$\text{(7.15)} \quad \begin{array}{ll} \text{blue} & : Or^+(x) \in A(a_2), \\ \text{red} & : Or^+(x) \in A(a_3), \\ \text{green} & : Or^+(x) \in A(a_4). \end{array}$$

The Map 69 shows the distribution of these colors on a sufficiently fine grid in $L \times [0.002, 1.998]$, where $h$ increases from top to bottom. Apparently on $L$ the basins of $a_2$, $a_3$ and $a_4$ compete for each $h$ and create a Cantor set-like distribution. Map 70 and 88 are based on the same experiment including information about level sets of equal dynamic distance to the centers (see (6.13)).

Maps 71–73 are derived from the same data base as Map 23 and illustrate the striking changes one obtains in graphical experiments through different color maps. For further information on the Julia-like sets and other findings, as for example the existence of strange attractors for (7.2), we refer to [PPS].

Without going into any details we finally comment on the Maps 55–57 and Map 74. They exhibit the data base of the same experiment using different color maps. Here $f(s) = s - s^3$. While the previous experiments were based on a two-meshpoint discretization of our boundary value problem (7.11) the discretization here is based on six interior meshpoints. Thus Newton's method (7.2) is now in $\mathbb{R}^6$, and our choice of parameter $\mu$ is such that $H(x) = 0$ has now five competing solutions. Our experiments show the intersection of the respective basins of attraction with some 2-dimensional hyperplane in $\mathbb{R}^6$.

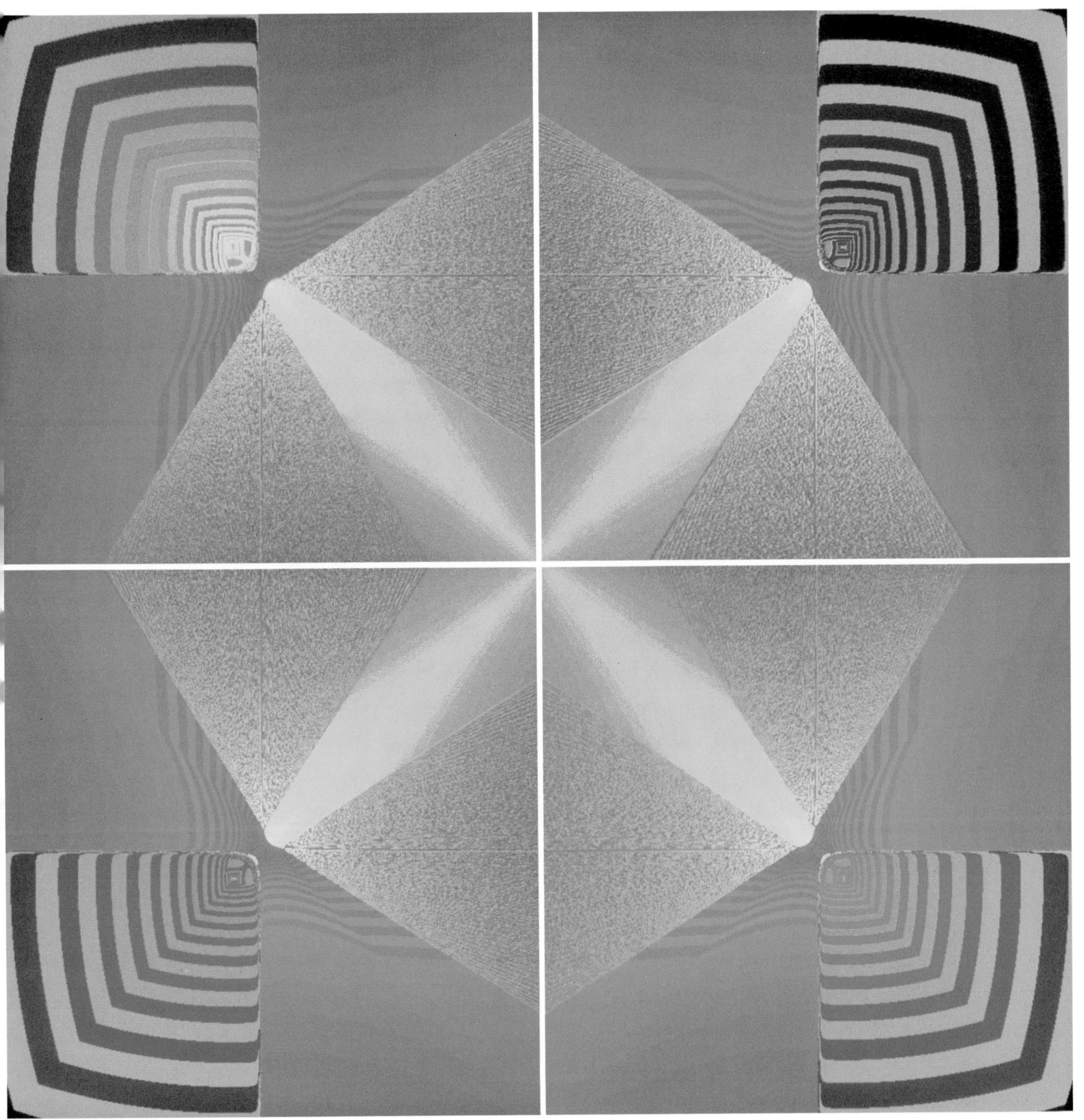

Map 68

Map 69

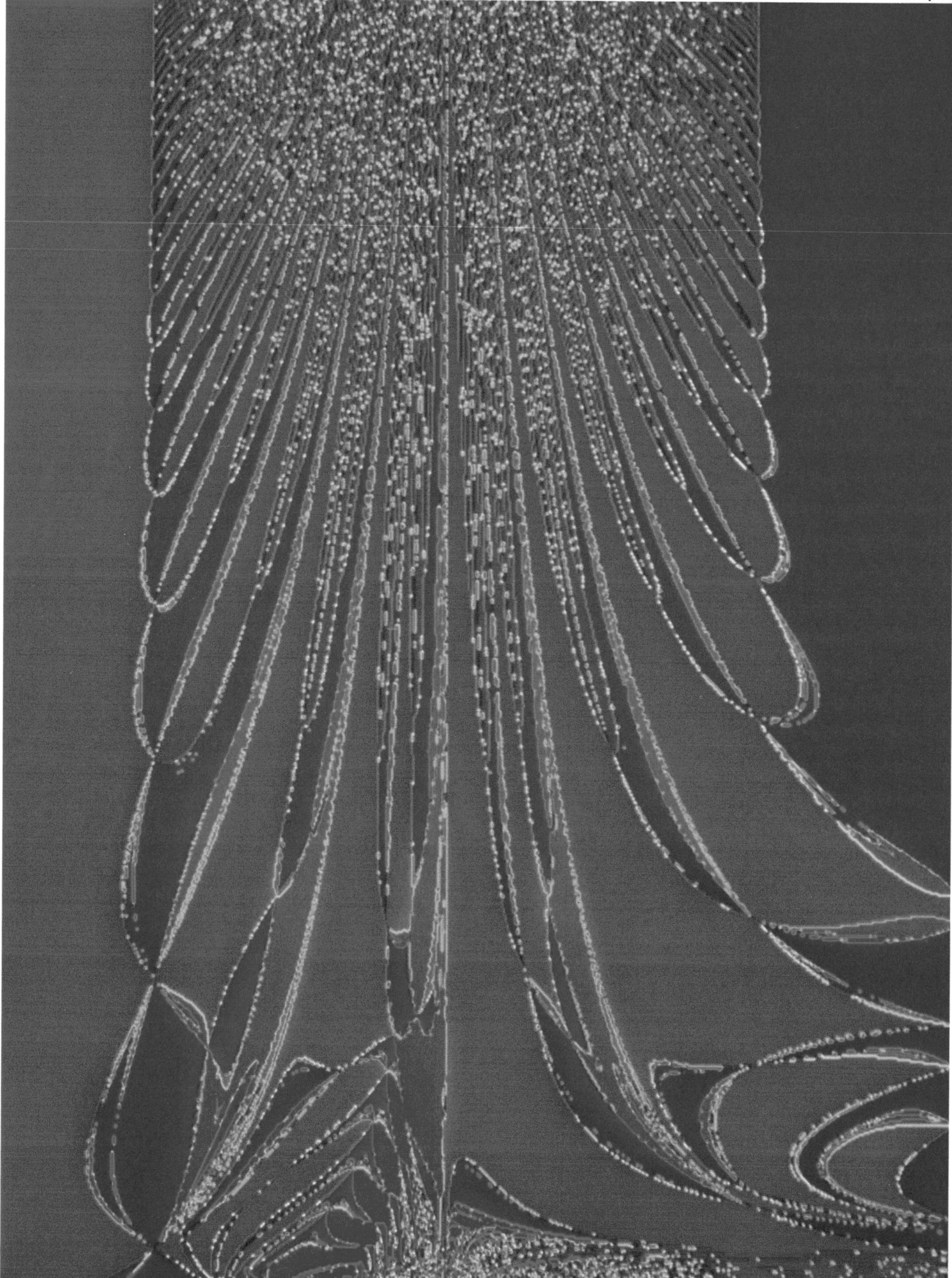

Map 70

Map 71

Map 72

▽ Map 73

Map 74

Map 75

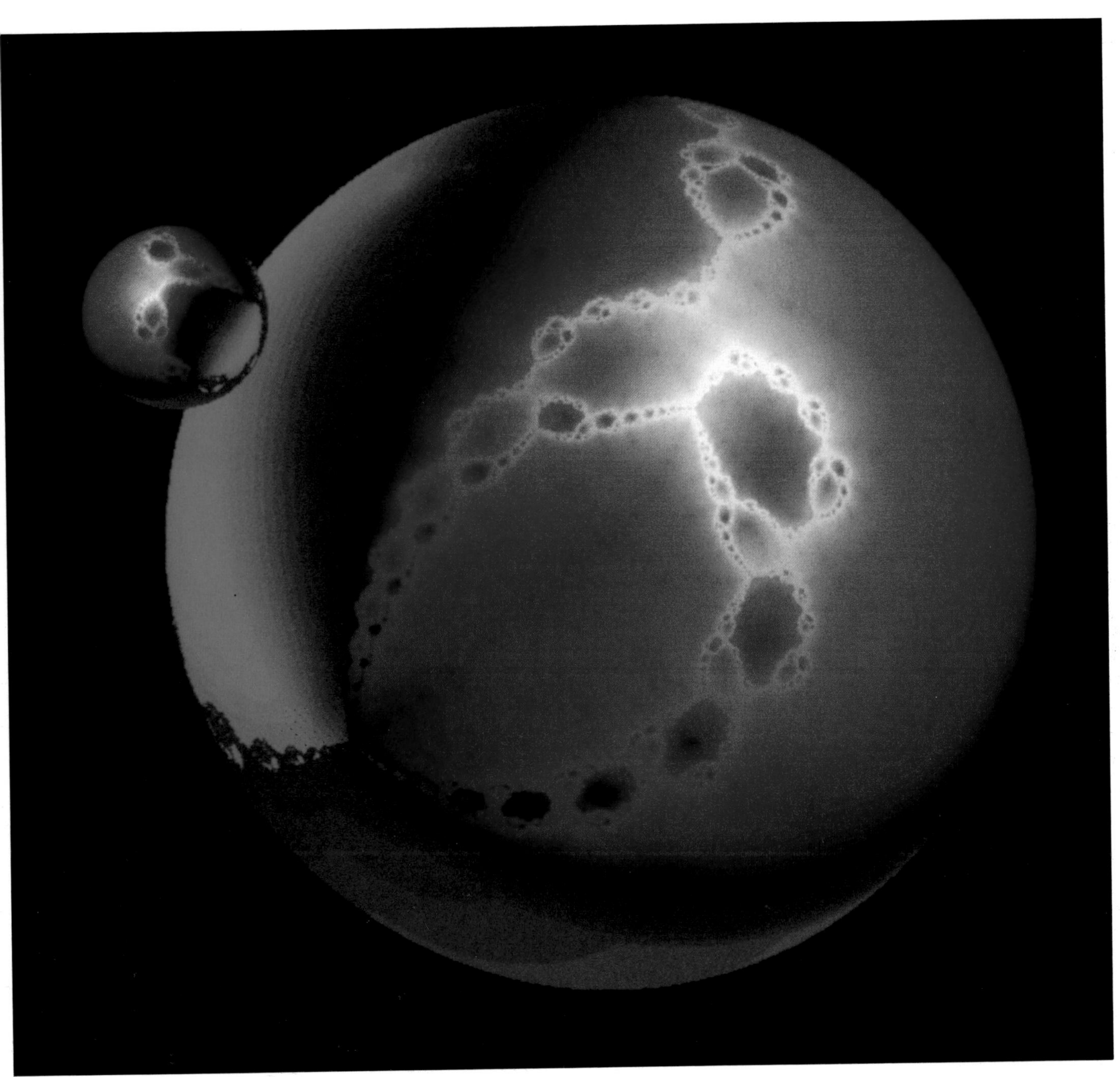

Map 76

Map 77

Map 78

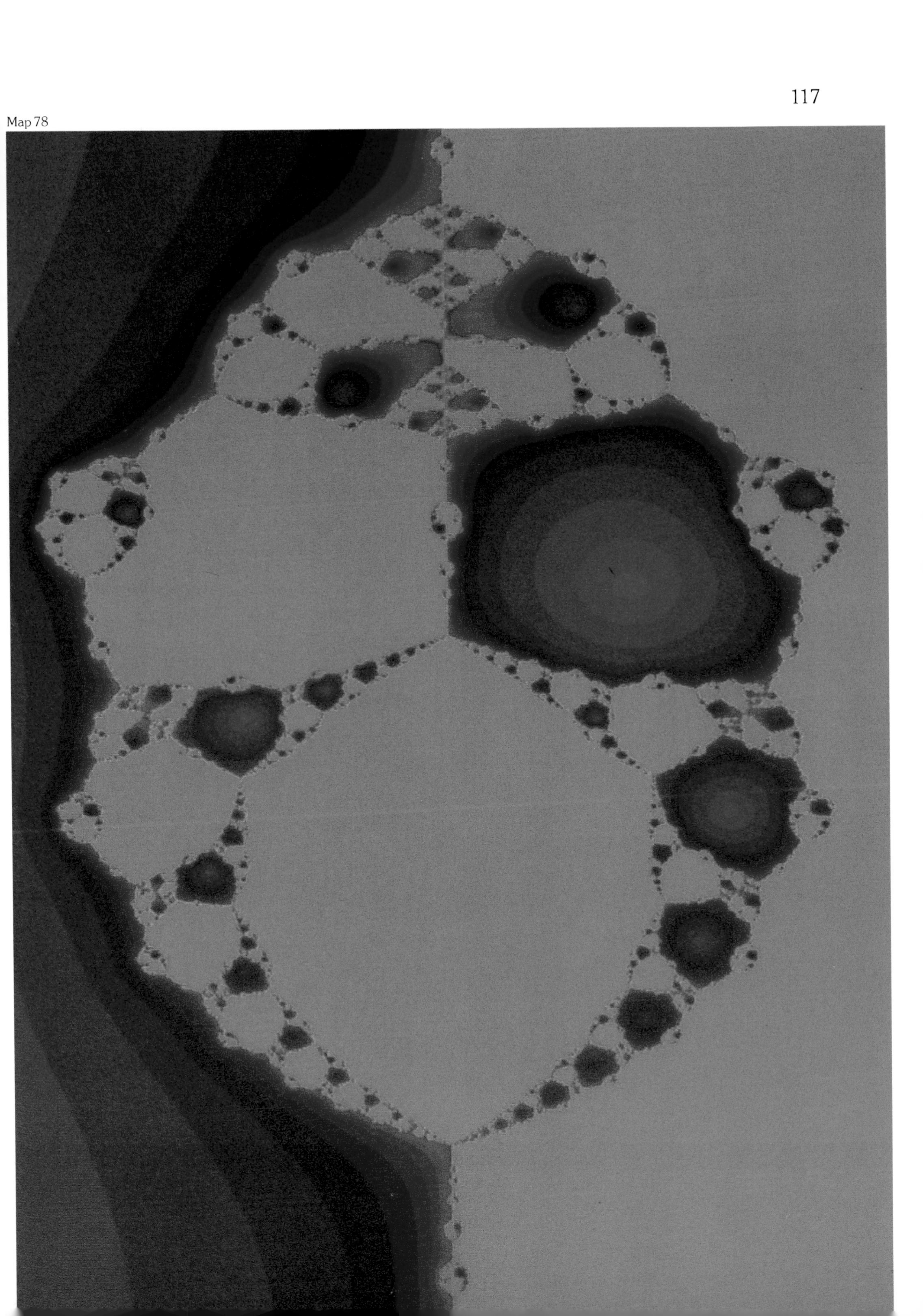

Map 79

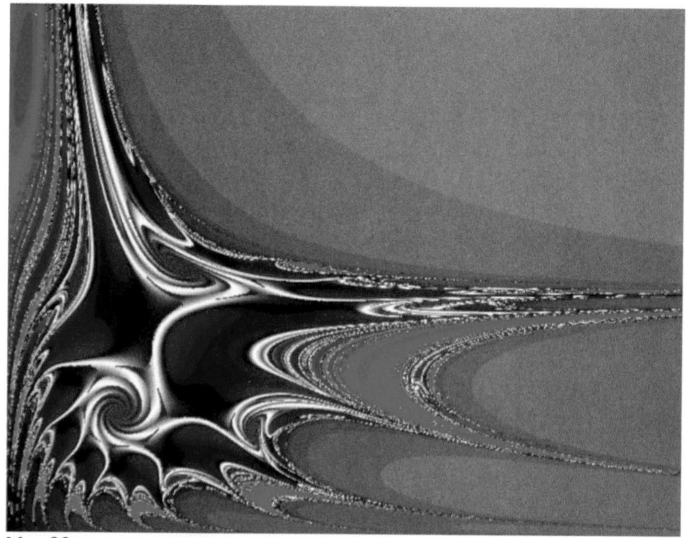
Map 80

Map 82

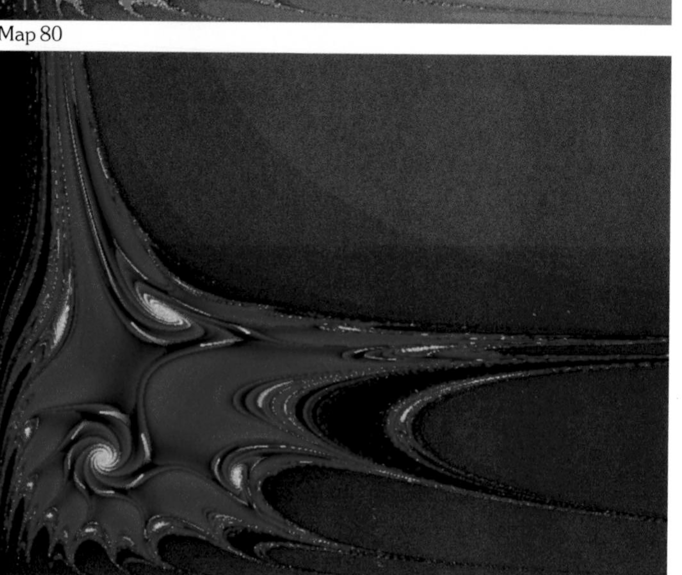
Map 81

Map 84

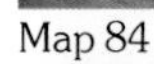

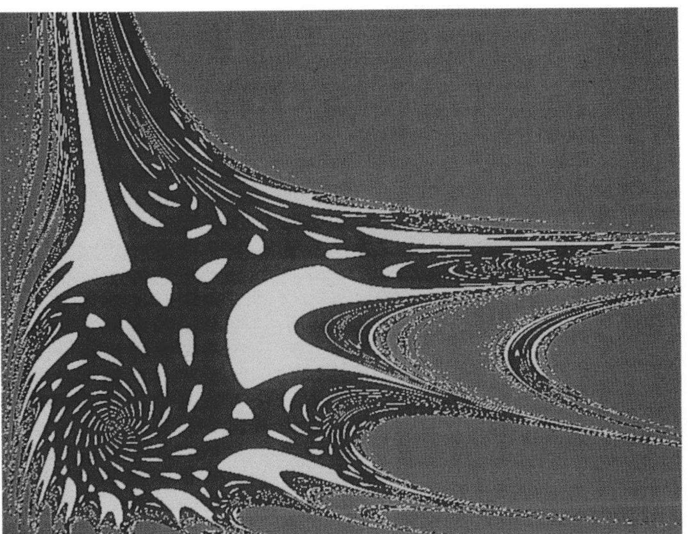
Map 83

Map 85

Map 86

Map 87

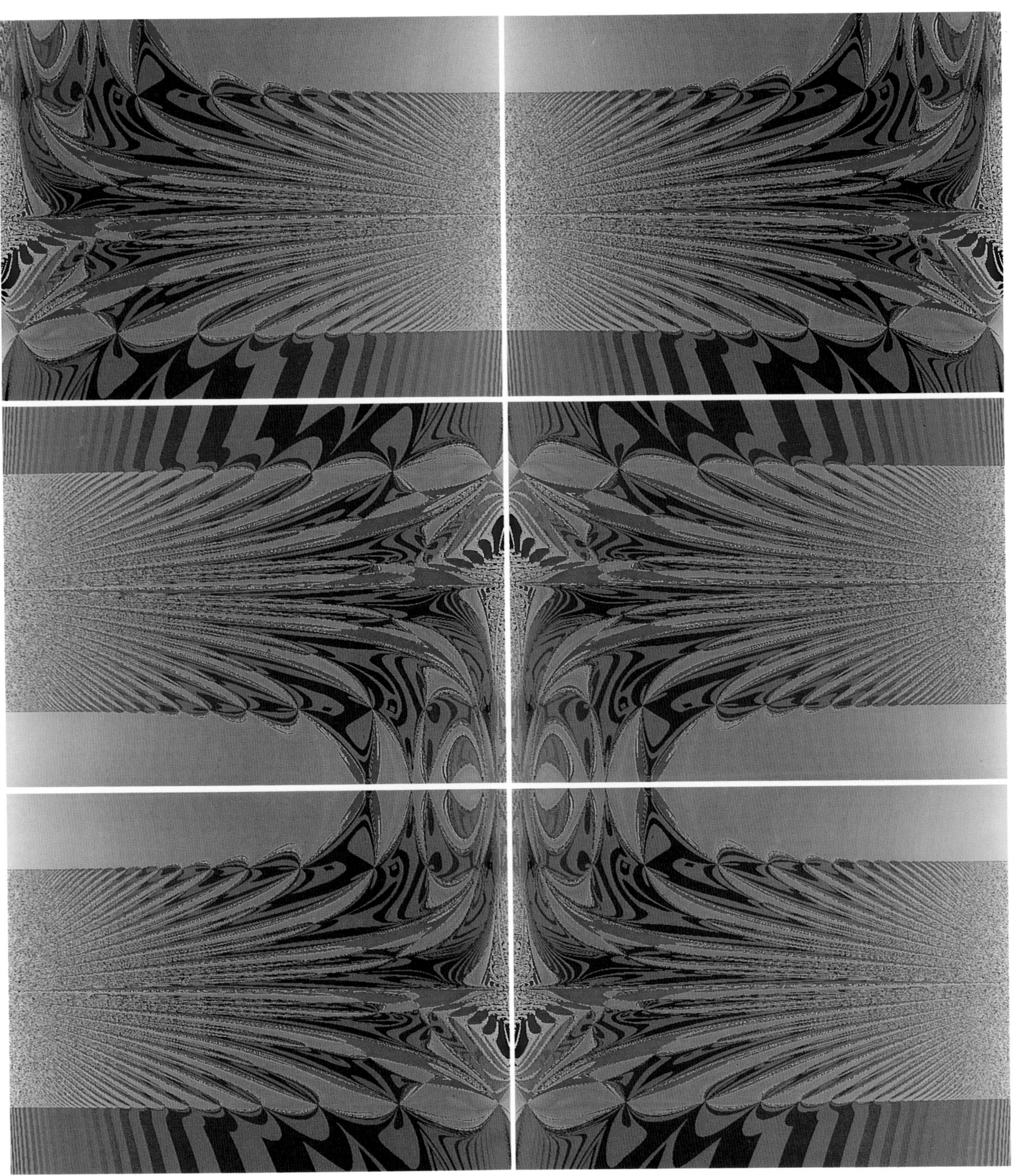

Map 88

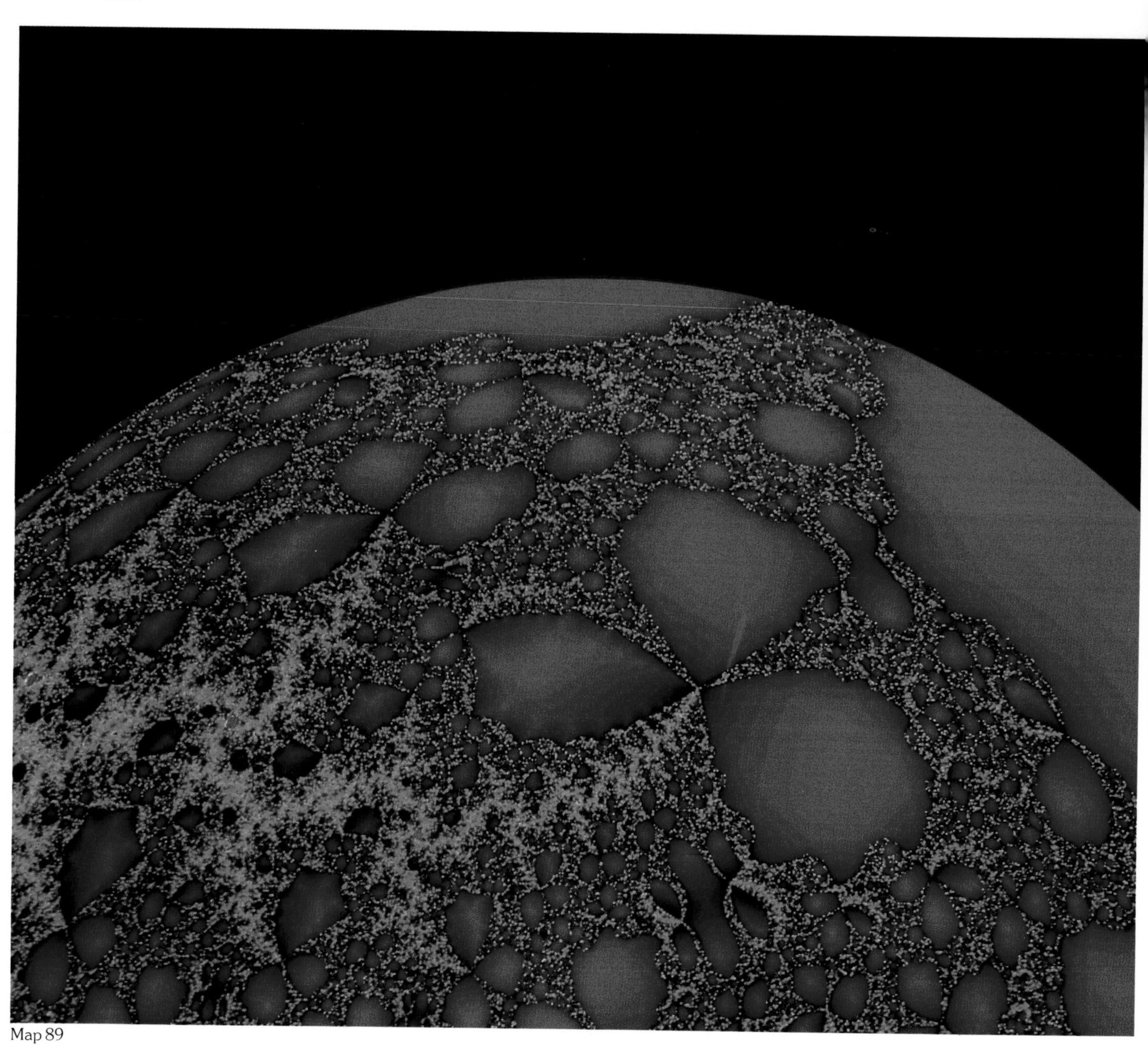

Map 89

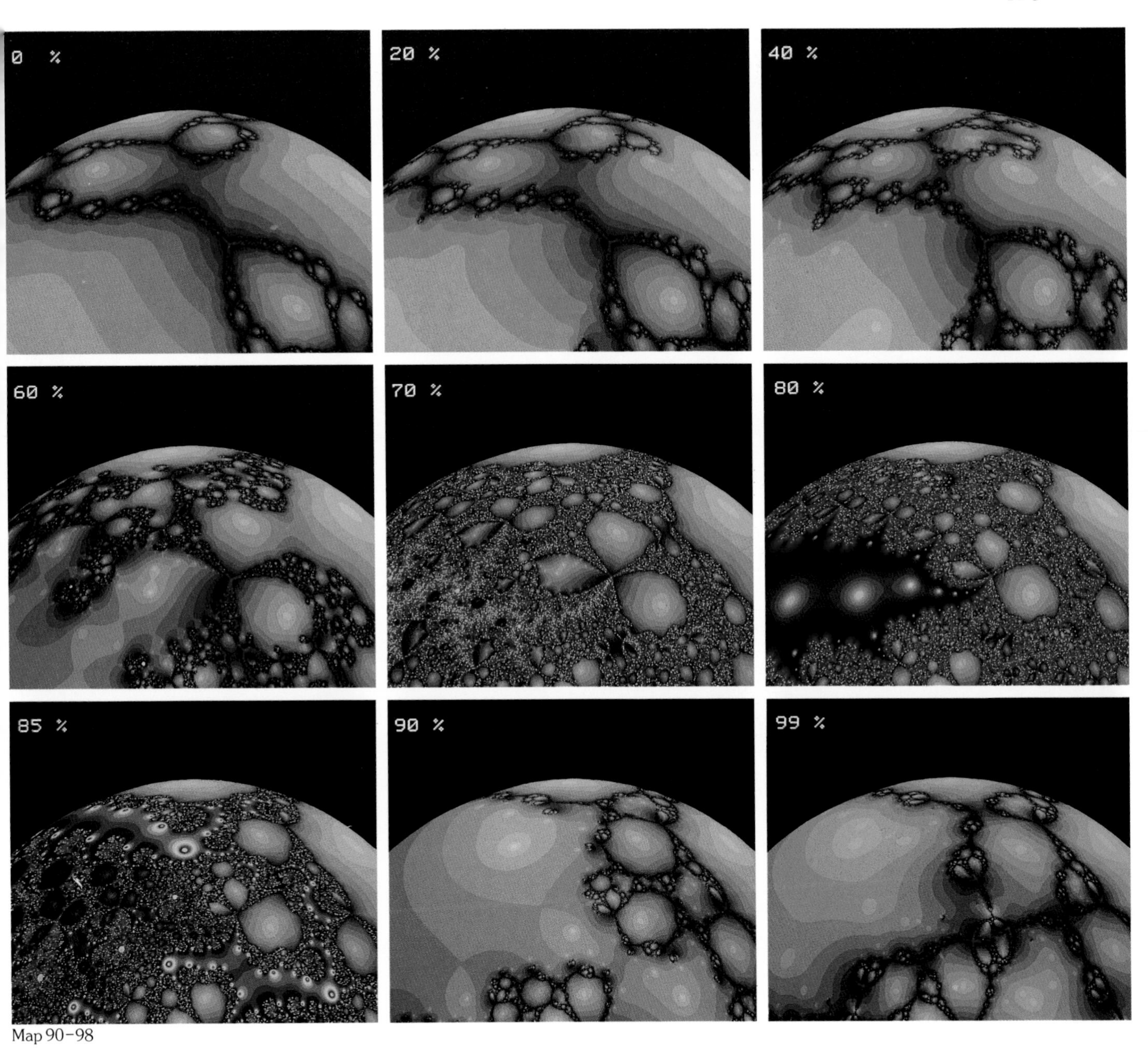


Map 90–98

Map 99

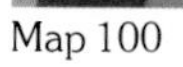

Map 100

Map 101

# 8 A Discrete Volterra-Lotka System

In our discussions of Newton's method we demonstrated how an apparently innocent system of differential equations gives rise to unimaginably rich and complex behaviour after discretization. This is also the Leitmotiv of our last example, the Volterra-Lotka equations for a predator-prey system.
Let $x(t)$ (resp. $y(t)$) be the population size of the prey (resp. predators) at time $t$. Then assume for the growth rates of $x$ and $y$:

$$(8.1)\quad \begin{cases} \dot{x}/x = \alpha - \beta y, \\ \dot{y}/y = -\gamma + \delta x, \quad \alpha, \beta, \gamma, \delta > 0. \end{cases}$$

Hence, $x$ grows at a constant rate in the absence of $y$, and $y$ decays at a constant rate in the absence of $x$. The prey is consumed in proportion to $y$, and the predators expand in proportion to $x$. System (8.1) rewritten is:

$$(8.2)\quad \begin{cases} \dot{x} = \alpha x - \beta xy = f(x, y), \\ \dot{y} = -\gamma y + \delta xy = g(x, y). \end{cases}$$

Obviously the point $(x_s, y_s) = (\gamma/\delta,\ \alpha/\beta)$ is a restpoint for (8.2), and it is well known that the first quadrant $\{(x,y): x>0, y>0\}$ is foliated by closed orbits centered around $(x_s, y_s)$, i.e. the populations experience periodic oscillations.
There are numerous numerical techniques to discretize (8.2), one-step methods, multi-step methods, etc. Our interest will be in a *mating* between two fundamental one-step methods: the Euler method and the Heun method.

$$(8.3)\quad \begin{cases} \begin{pmatrix} x_{k+1} \\ y_{k+1} \end{pmatrix} = \Phi_{h,\rho} \begin{pmatrix} x_k \\ y_k \end{pmatrix}, \quad k = 0, 1, \ldots \\ \Phi_{h,\rho} \begin{pmatrix} x \\ y \end{pmatrix} = \begin{pmatrix} x \\ y \end{pmatrix} + \dfrac{h}{2} \begin{pmatrix} f(x,y) + f(x+\rho f(x,y), y+\rho g(x,y)) \\ g(x,y) + g(x+\rho f(x,y), y+\rho g(x,y)) \end{pmatrix} \end{cases}$$

Note that $\rho = 0$ gives Euler's method, while $\rho = h$ yields Heun's method.
Henceforth we let $\alpha = \beta = \gamma = \delta = 1$ and observe that

$$\Phi_{h,\rho} \begin{pmatrix} 1 \\ 1 \end{pmatrix} = \begin{pmatrix} 1 \\ 1 \end{pmatrix}$$

for all $h$ and $\rho$.

To discuss (8.3) we look into the $(h, \rho)$ – parameter plane in Fig. 51. Qualitatively, we can distinguish four domains in that plane: For $(h, \rho)$ in (I), the fixed point $(x, y) = (1, 1)$ is attractive. As we change $(h, \rho)$ and move from (I) into (II), system (8.3) experiences a Hopf bifurcation, i.e. the fixed point becomes unstable and gives birth to an *invariant circle* which is attractive. This characterizes $(h, \rho)$ in (II) except for choices from the little *tongues* emanating from the boundary of (I). If $(h, \rho)$ is from one of the infinitely many tongues, there is an attractive periodic orbit (consisting of a finite number of points) instead of the invariant circle. One says the system is in *resonance*.

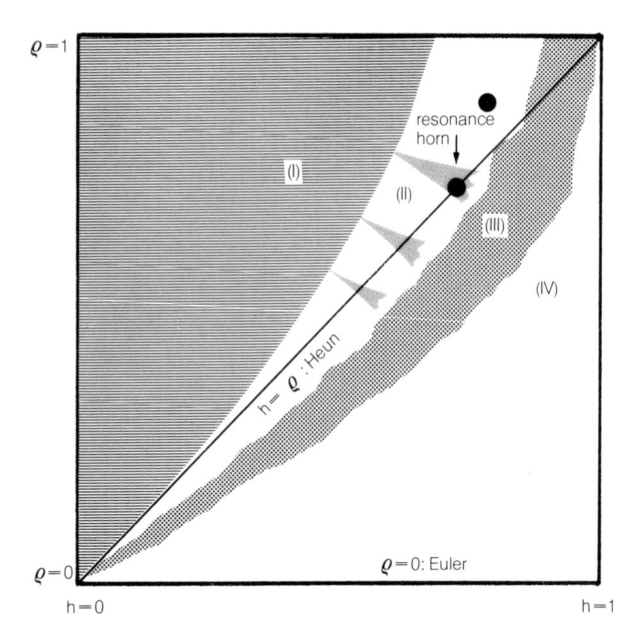


*Fig. 51. The (h, ρ)-parameter plane*

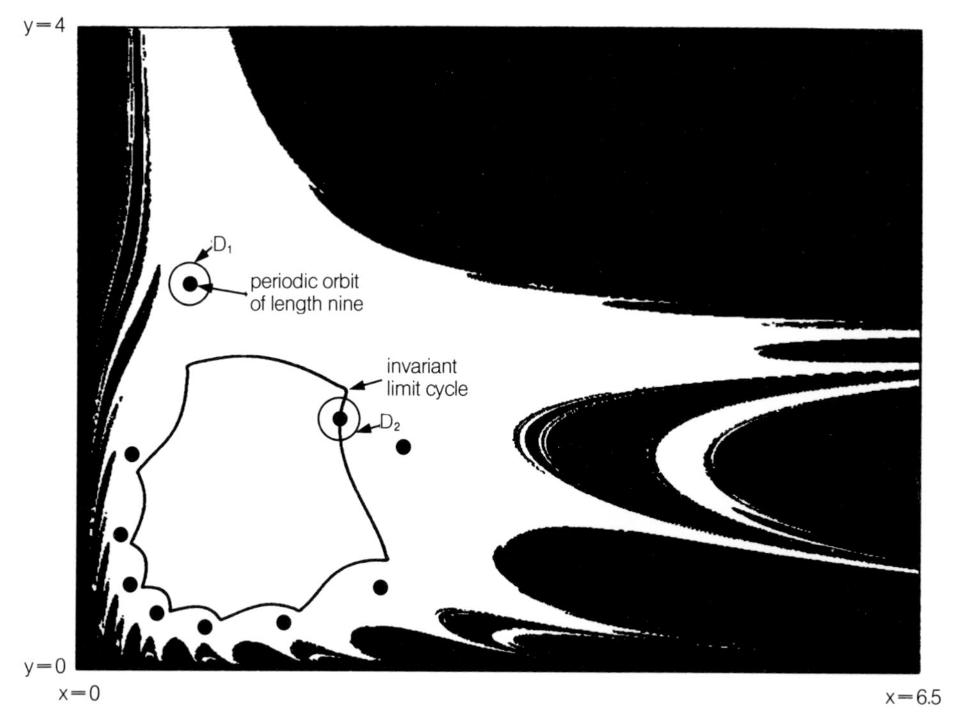


*Fig. 52. Target sets ($D_1$ and $D_2$) for decomposition into level sets (see also Maps 83–86)*

If one moves $(h,\rho)$ further, the system exhibits *strange attractors* (III) and eventually in (IV) $\Phi_{h,\rho}$ admits no finite attractor, i.e. all orbits go to $\infty$, except of course for the point (1, 1) and its preimages and possibly some other repelling periodic points.

Figure 51 has to be regarded with some caution. It is just a very rough sketch of prominent qualitative behavior. The mechanisms, for example, which cause the transition from (II) into (III) are far from being understood. Experimentally, however, if $(h, \rho)$ is in (II) and "close to (III)" we typically found two disjoint attractive orbits, such as for example, an attractive invariant circle coexisting with an attractive periodic orbit.

Maps 83–86 illustrate such a possibility. There $\rho=h=0.739$ and we have a situation as sketched in Fig. 52.

In this case an invariant circle coexists with an orbit of period 9. Map 83 shows the basins of attraction: yellow for the periodic orbit, red for the invariant circle, and grey for all initial points whose orbits go to infinity. Applying a decomposition into *level sets of equal attraction* from the attractors (see Special Section 2) we obtain Maps 84–86. As target sets $L_0$ we have chosen a disk like $D_1$ (see Fig. 52) for the periodic orbit and a disk like $D_2$ centered at an arbitrary point on the invariant circle (see Fig. 52).

In Maps 79–82 we have $h=0.8$ and $\rho=0.86$ which is a point in (II), illustrating the situation after a Hopf bifurcation, i.e. we have just one attractive invariant circle. Map 82 shows its basin of attraction in red, while initial points from the grey region belong to orbits which go to infinity. Maps 79–81 show again a decomposition into *level sets* which are obtained in the same way as before.

# MAGNETISM AND COMPLEX BOUNDARIES

In spite of all their differences, the various branches of science regularly display similarities in their basic thought structures. The reservoir of concepts and methods reflects the common cultural background of scientists more than the differences between the individual fields of study. Experiencing this unity is deeply gratifying; it is a reward of scientific endeavor, for when new schemes are developed for the solution of one *specific* problem, they unexpectedly throw a new and productive light on quite different questions. Fascinated by this interchange of ideas, K.C. Cole, in her book *Sympathetic Vibrations,* cites a dictum from V.F. Weisskopf, "What's beautiful in science is that same thing that's beautiful in Beethoven. There's a fog of events and suddenly you see a connection. It expresses a complex of human concerns that goes deeply to you, that connects things that were always in you that were never put together before".

*Self-similarity* is actually an old idea. Russian dolls, for example, embody the belief in self-similar homunculi, all originally contained in Eve's ovaries as succeeding generations of humankind. In the 18th century, A. von Haller and others developed this belief into a respected theory of preformation. Their historical defeat in the dispute with the epigeneticists may have been partly responsible for self-similar structures being ignored until around 1970, when they were re-introduced with a new enthusiasm. Independently, although almost at the same time, K.G. Wilson in physics (renormalization theory) and B.B. Mandelbrot in mathematics (fractal geometry) reaffirmed the existence of this kind of structure, thereby creating new instruments that greatly enrich our repertoire of thought patterns.

The close relationship of these developments is documented in our pictures from the theory of magnetism. In these pictures, the physics of phase transitions touches the mathematics of Julia sets. At the present time, it is not known whether this is a mere coincidence or actually the reflection of a basic principle of phase transitions. Summing up his history of metallurgy, Cyril S. Smith said "... discovery derives from aesthetically motivated curiosity and is rarely a result of practical purposefulness". We hope that the following descriptions will motivate further studies. Fractal structure of phase boundaries may well turn out to be typical. It will then be a challenge to explore the physical meaning of this mathematical characterization.

## *Phase Transitions and Renormalization*

*Phase transition* is defined as the change between states of matter. In school we all learned the three basic states *solid, liquid,* and *gas,* but there are numerous other states, or *phases,* that can be identified if we look more closely. Many solids change their crystalline structure when temperature or pressure are changed. At very high temperatures or very low densities matter becomes ionized and, as *plasma,* has properties which are very unusual on earth but common in outer space.

The most intensively studied phase change is the transition between the *magnetic* and *non-magnetic* states of a material. Many substances possess elementary atomic magnets that have the distinct tendency to line up in parallel.

As long as the thermal fluctuations are not too strong, this tendency leads to the macroscopic, observable order which we know as magnetism. This order progressively becomes more diffuse as the temperature is raised, and order turns into disorder in a characteristic way at the so-called *Curie temperature* (named after Pierre Curie, the husband of Marie Curie). For iron, this happens at 770 °C. Above this temperature there is a hint of magnetism: within certain distances and for certain time spans the elementary magnets can hold together, but they can no longer "communicate" over large distances and for long times. The higher the temperature, the smaller the typical lengths and times of coherence, until at very high temperatures the elementary magnets change their direction completely independently.
Analogously, other phase transitions can also be described as either the formation or the destruction of *long-range order,* with the same connection between temperature and space-time scale of coherence outlined above. The observation that very different systems – for example, a magnet near its Curie temperature and a liquid at its critical point – are surprisingly similar in a quantitative way, was both amazing and, until the end of the 1960's incomprehensible. The microscopic nature of the order appeared to be irrelevant to the understanding of the phenomena. But then what was the basis for the observed universality?
The decisive idea came from L. P. Kadanoff in 1966, while he was working at Brown University. Without explicitly using the term self-similarity, he developed the basic concept it expresses. He considered how the same magnet would look at different scales. To begin with, take a magnet at absolute zero, where all the elementary magnets are lined up. Its appearance does not change as we vary the scale. Whether our observation "window" extends from 1 to 100 nanometers or from 1 to 100 micrometers, the *order is complete.* One can take the smallest perceived unit (a nanometer- or a micrometer-block, respectively) as the elementary magnet and find that, at absolute zero, they are all lined up perfectly.
The scaling behavior is similar when the magnet is at "infinitely" high temperature. There the atomic magnets all fluctuate totally independent of one another; at all scales above the atomic there is *complete disorder,* so that the picture in the nanometer range is again the same as in the micrometer range.
What happens when the temperature is neither zero nor infinite? Let us first consider low temperatures. There is indeed a macroscopic order, but it is not complete, since some of the atomic magnets get out of line because of thermal fluctuations. If we compare the various scales, we do see differences. For example, the fluctuations may extend to the nanometer range, but not reach farther. At micrometer range they would be imperceptible; there the magnet just looks as if it had the temperature zero. A coarsening of the scale from nanometers to micrometers thus corresponds to an effective lowering of the temperature. Now consider a magnet at high temperatures. Here, disorder is not complete. There are always local aggregates of lined-up atomic magnets. But again, coarsening of the scale – a *scale transformation* – causes the small coherent regions to disappear. When the scale is coarse enough, the magnet appears to have an infinitely high temperature.

This strategy is reminiscent of decision making. In order to assess a complicated situation, it is often helpful to view it from a progressively more distant point. The picture becomes clearer as details are averaged out.
The essence of this idea is to relate scale transformations to changes in temperature. The same magnet of given temperature, when viewed on different scales, looks as if it were at different temperatures. We say a scale transformation forces a corresponding *renormalization* of temperature.
Consider a magnet of $N$ atoms with inter-atomic distance $a$ and temperature $T$. On a coarser scale where the elementary block is taken to have sidelength $a' = b \cdot a$ and comprises $b^3$ atoms, the magnet looks like one with $N' = N/b^3$ atoms but with another, *renormalized* temperature $T'$. The relation $T' = R_b(T)$ is called *renormalization transformation.*
Our qualitative discussion has shown that this transformation has two *attractors:* the temperatures $T=0$ and $T=\infty$. This is because magnets at low temperatures yield lower and lower renormalized temperatures for successively coarser scales, while high temperatures are renormalized even higher. We say that low temperatures are in the domain of attraction of $T=0$, and high temperatures are in the domain of attraction of $T=\infty$. The Curie temperature $T_c$ is the boundary between these two domains of attraction. When a magnet is at that temperature, it looks the same on all scales. Its temperature does not change under renormalization, $R_b(T_c) = T_c$, simply because it cannot decide to which attractor it should go. In the language of dynamical systems we say that $T_c$ is a *repeller* for the renormalization process. If a magnet's temperature deviates ever so slightly from $T_c$, that deviation gets amplified by renormalization, and iteration of the process leads to one of the clear cases, i.e. to complete order ($T=0$) or complete disorder ($T=\infty$).
We imagine the magnet at the temperature $T=T_c$ to have coherent fluctuations of all ranges woven together, small fluctuations imbedded in larger ones and so forth. Briefly, the pattern of fluctuations at the critical value of the temperature is *self-similar.*
This basic idea eventually led to quantitative results and explained the physics of phase transitions in a satisfying way. After all, the way from the idea of renormalization to its concrete, final form was so elusive that Kadanoff did not find it. Rather, K.G. Wilson at Cornell in 1970 surmounted the difficulties and developed the method of renormalization into a technical instrument that has proven its worth in innumerable applications.
The growth of the renormalization idea from Kadanoff's vision to Wilson's practical method had an intriguing side effect: the suggestive picture of self-similar fluctuations at the critical point faded into the background while the method took on a very technical and less comprehensible form. The analysis became occupied primarily with the question of how quickly the renormalization moved away from the boundary region around the critical point. It seems amazing that the explanation of experimental results concerning the phenomenology of phase transitions did not need to deal explicitly with self-similar structures.
Renormalization theory, however, has recently led to fractal phase boundaries in an unexpected way so that it bears, even though now unrelated to experimental questions, an intuitively appealing relationship to Kadanoff's

idea of self-similarity. In order to explain this, we have to go a way back into history.

### *Yang and Lee's Zeros*

In 1952, the physicists C.N. Yang and T.D. Lee suggested extending the treatment of phase transitions from the real world into the mathematical world of *complex numbers.* This may seem arbitrary and inapplicable from the standpoint of physics, since an imaginary temperature or magnetic field does not seem to make sense. Nevertheless, this was not the first or last time that a look beyond the curtain of reality gave a definite impulse to our basic understanding. Think of Gauss's elegant solution of the problem of the zeros of a polynomial. The equation

(1) $c_0 + c_1x + c_2x^2 + \ldots + c_Nx^N = 0$

has exactly *$N$ zeros in the complex* $x$-plane, while there is not such a simple, fundamental theorem for real $x$. Or remember Mandelbrot's extension of the process $x \mapsto x^2 + c$ into the complex plane. The analysis of this and other iteration processes was made more comprehensive just because Fatou's theorem on the behavior of critical points applies only in the complex plane.
Yang and Lee posed the basic question of how the formalism of canonical statistical mechanics, established by Boltzmann and Gibbs, could describe transitions. There was an irritating discrepancy between the harmless-looking canonical recipe for the calculation of the thermodynamic quantities and the necessity for these quantities to become singular at phase transitions. The problem was to clarify the mathematical nature of these singularities. It appeared – according to the theory – that these singularities should not exist in the real world.
For a large class of physical systems, Yang and Lee could rephrase the question as the problem of finding and characterizing zeros of a certain function. The desired singular points – candidates for the phase transition – were solutions of a polynomial equation of the type (1). But because all the coefficients $c_0, c_1, \ldots, c_N$ turned out to be positive, there could be no solutions in the region of physically realizable positive $x$. What was more natural than to look into the complex plane where, according to Gauss, $N$ solutions must exist? It could be, so went the idea, that the complex zeros accumulated close to the real $x$-axis and thus identified the desired singularity. This would be made possible by having the transition to real physics be made in the limit $N \to \infty$, because it was known that "true" phase transitions only occur in the limit of very large numbers of particles.
Yang and Lee finally developed the following picture. For a finite number of particles, the polynomial equation (1) yielded a finite number of zeros in the complex plane. As the number of particles increased, the number of zeros had to increase. The set of zeros had to get denser and, at the same time, to come closer to the real $x$-axis. In the limit $N \to \infty$ there was the possibility that the infinite set of zeros condensed to a line that intersected the real $x$-axis at the point of the physical phase transition $x = x_c$. The real regions $x < x_c$ and

$x > x_c$, which characterized the phases 1 and 2, obtained a natural extension into the complex plane, and the line of zeros could be interpreted as a *complex phase boundary.*
In their famous theorem, Yang and Lee showed in a mathematically exact way that this picture holds for the behavior of many magnetic materials in an externally applied magnetic field. The quantity $x$ in polynomial (1) represents a complex magnetic field. If one is interested in the behavior of the magnet as a function of temperature, however, the problem is more difficult. In 1964 M.E. Fisher was able to show that the Yang-Lee picture with a complex temperature can be applied to the special case of a two-dimensional, quadratic lattice of elementary magnets with isotropic interaction between nearest neighbors. He showed that the zeros condense to a smooth line in the thermodynamic limit $N \to \infty$.
Toward the end of the sixties this result encouraged a number of physicists to relate quantitative aspects of phase transitions to the properties of the Yang-Lee zeros (S. Großmann, 1969). It was generally assumed that the zeros condense to a line in the limit $N \to \infty$. That this is not true in all cases would have been noticed sooner but for the spectacular success of renormalization theory [Wi], which absorbed the attention of theoreticians working on phase transitions. Not until 1983/84 was it realized that the picture of the Yang-Lee distribution of zeros was not so simple. Following up the twenty year old work of M.E. Fisher, it was shown [SC, SK] that for anisotropic two-dimensional lattices there can be two-dimensional distributions of zeros. Shortly before this, B. Derrida, L. DeSeze, and C. Itzykson found *fractal patterns* in so-called *hierarchical* lattices. The particular attraction of this discovery was that it used renormalization theory in an essential way, described in detail in the following section.

### *Fractal Phase Boundaries and Their Morphology*

The series of pictures in Fig. 53 shows a metamorphosis of complex boundaries. These come from a model of magnetism in which a selectable material parameter $q$ is varied slightly from picture to picture. The details of the model are explained in the Special Section 10. It should be noted that this is not meant to be a realistic description of magnetism. Rather the model has been used as an exactly soluble case to clarify basic issues in renormalization theory.
One such basic issue is the connection between the ideas of Yang and Lee on the one hand and Kadanoff and Wilson on the other. Can the picture of a phase boundary consisting of singularities be united with the idea of a boundary between two or more domains of attraction? This is precisely what the so-called *hierarchical models* achieve in a mathematically exact way and herein lies their theoretical value. For these models it can be shown that the *Julia set of the renormalization transformation is identical with the Yang-Lee set of zeros.* This is explained more fully in the Special Section 10. That Julia sets are important in this context comes from the lucky circumstance that the renormalization transformation happens to be a rational mapping. There is reason, however, to expect that the idea of Julia sets can be carried over to

*Fig. 53. Yang-Lee singularities as Julia sets of the renormalization transformation (10.1). The window shown in the x-plane is for* a, b, *and* c: $-6 \leq Re\, x \leq 4$, $-5 \leq Im\, x \leq 5$; *for* d–k: $-5 \leq Re\, x \leq 5$, $-5 \leq Im\, x \leq 5$; *for* l: $-4 \leq Re\, x \leq 6$, $-5 \leq Im\, x \leq 5$

non-rational mappings, as R. Devaney has shown in his studies of the iteration of transcendental mappings, resulting in most beautiful pictures.
The structures in Fig. 53 are phase boundaries in the sense of Yang-Lee and in the sense of renormalization theory, extended into the complex plane. Their most conspicuous common feature is their self-similarity. Could this be an accidental by-product of the artificial hierarchical lattice structure, or does it reflect a physically real self-similarity at the phase transition? Presently, this question is an open issue. If it were to be verified that the complex phase boundaries are typically fractal structures, then the fractal dimension and other characteristics of fractals ought to be physically meaningful quantities.
Another common property of the structures in Fig. 53 is that they all have one and only one accumulation point $x_c$ on that part of the real axis which describes ferromagnetic substances at positive temperatures: $1 < x < \infty$. That feature corroborates well the original picture proposed by Yang and Lee even though the singularities fail to lie on smooth lines.

*Fig. 53* (continued)

The interested reader can find the exact definition of the variable $x$, and of a material parameter $q$, in the accompanying Special Section 9. The value $x=1$ corresponds to infinite temperature, i.e. complete disorder, and $x=\infty$ represents absolute zero with complete ferromagnetic order. The point $x_c$ corresponds to the Curie temperature; it separates the ferromagnetic phase from the paramagnetic phase for real temperatures. Its complex extension is the Julia set, which clearly separates two phases for, say, $q>3$ and $q\leq 0$ (magnetic phase outside, non-magnetic phase inside, see Fig. 53). In contrast, for $0<q\leq 3$, the zeros in the thermodynamic limit form structures which separate infinitely many regions from one another. Does this correspond to so many "phases", or are there still only the two attractors $x=1$ and $x=\infty$? Computer experiments, such as shown in Map 3–6, give us the answer. The colors here identify the different regions of attraction; their gradation represents the dynamic distance from the corresponding attractor with respect to the iteration of the renormalization transformation. It turns out that for $1.21<q\leq 3$ just two phases are interwoven. For $0<q<1.21$ there is a third

phase, which we may call antiferromagnetic, because it includes the point $x=0$ ($J<0$, $T\to 0$ see Special Section 9); for $q=1$ this is an attractive fixed point. Map 6 shows a detail in which the ferromagnetic phase is red, the paramagnetic phase green, and the antiferromagnetic phase grey. For comparison, Map 5 shows in enlarged detail that even two phases can weave themselves together to organic complexity. (The red color is used to accentuate the edge of the blue region.)
The color plates Map 7–10 are examples of complex phase diagrams from the second of the hierarchical models we investigated (see Special Section 10 for details). Map 7 shows the case $q=2$ (Ising spin). It shows three phases: the ferromagnetic in blue, the paramagnetic in white/grey, and the antiferromagnetic in red. At physical temperatures there are two phase transitions, a Curie point $x_c>1$ and a corresponding transition to antiferromagnetism at a so-called Néel point $x_c<1$. Maps 8–10 are magnified details showing how one phase can be sprinkled with an infinite number of isolated islands of another phase.
The multitude of structures in Fig. 53 and Maps 3–10 needs to be analyzed to reveal their ordering principles, just as the Julia sets of the processes $x \mapsto x^2+c$ were analyzed in Mandelbrot's studies. Here, again, the investigation of the critical points of the mapping holds the clue. The details of the argument are presented in Section 10. We find that in the first hierarchical model – just as with $x \mapsto x^2+c$ – the fate of the point $x=0$ is decisive, whereas in the second model the analysis involves also the point $x=1-q$.
Map 1 represents the morphology of the phase boundaries in the case of the first model. It shows the plane of all possible values of the model's material parameter q. For $q$-values in the green or yellow region there are two phases, for values in the black region three, and in some of the buds even four phases. If $q$ is picked outside of the peculiar cauliflower shaped structure, the phase boundary is a more or less crumpled circle (s. subpictures a, b, and l in Fig. 53). If $q$ is in the colored interior of the "cauliflower", the corresponding phase diagram has a similar structure; the magnetic phase reaches deep into the interior of the non-magnetic domain. The qualitative details of this structure are the same for q-values inside a given component of the cauliflower, but they vary from component to component. For yellow values of $q$, $x=0$ is in the domain of attraction of the ferromagnetic center $x=\infty$; for green q-values, $x=0$ belongs to the paramagnetic phase. The complex alternation of these tendencies in the neighborhood of the black region is bewildering, as is shown impressively in the enlarged sections in Maps 11–13. This becomes progressively tangled and takes on a random character in the immediate proximity of the black region, a detail of which is shown in Map 2 with a surprise: the well-known Mandelbrot figure appears; its identity with the original Mandelbrot set is astounding.
Figure 54 and Maps 14–16 show analogous results for the second hierarchical model. Again it turns out that the appearance of a further phase and the associated drastic changes in the nature of the boundary are governed by the Mandelbrot set.
Perhaps we should believe in magic. Even if one accepts that more general dynamical laws might look locally like $x \mapsto x^2+c$, it is still amazing that the

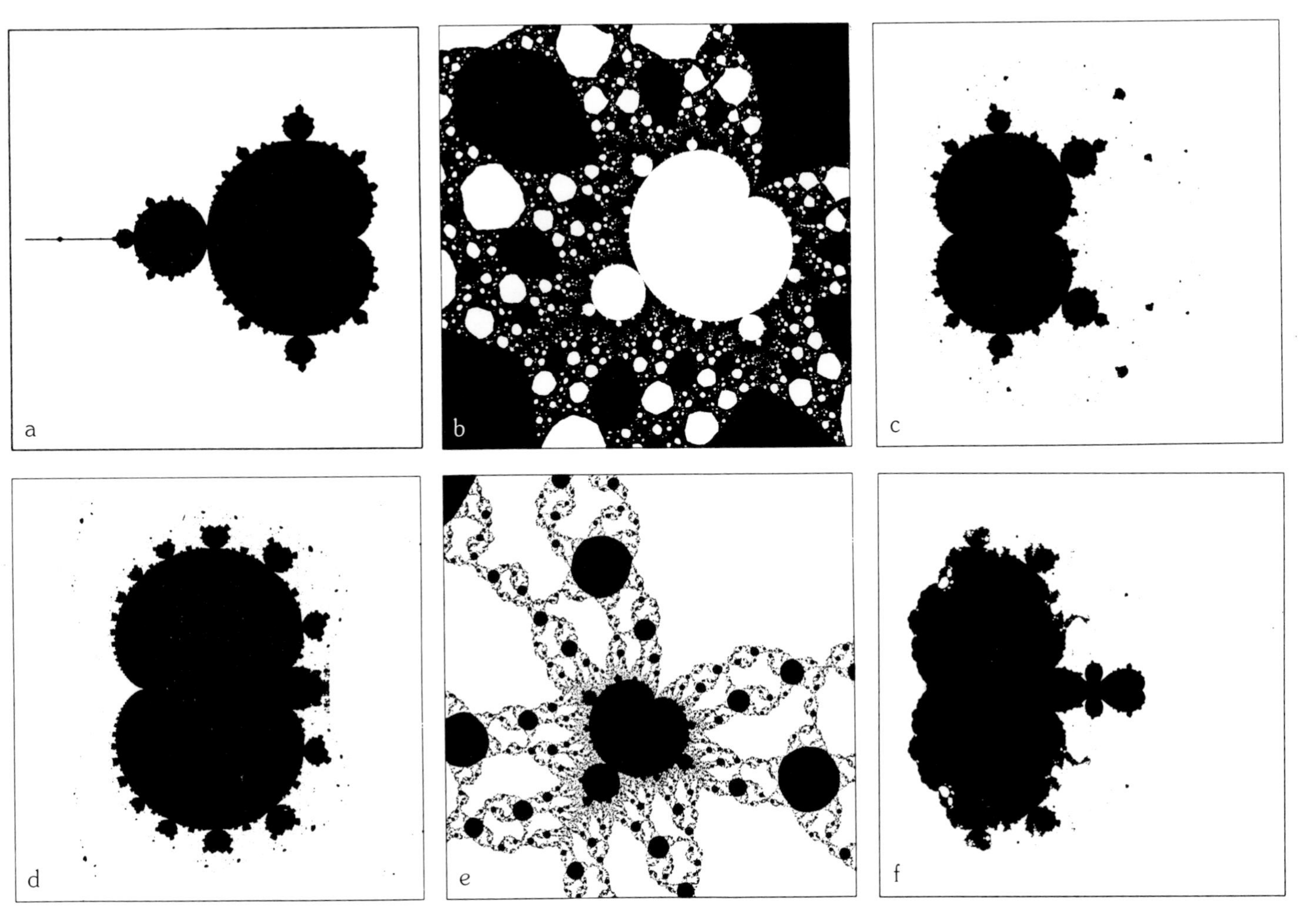


*Fig. 54 a–f.* Studies of critical points. a: *The Mandelbrot set for* $x \to x^2 + c$; b: *detail from* c; window: $1.85 \leq Re\ q \leq 2.15$, $1.5 \leq Im\ q \leq 1.8$; *for black values of q the orbit of* $x = 0$ *goes to* 1 *under renormalization.* c: *Hierarchical model as in Fig. 55;* window: $-1 \leq Re\ q \leq 3.5$, $-2.25 \leq Im\ q \leq 2.25$; *for black values of q, the orbit of* $x = 0$ *goes neither to 1 nor to* $\infty$. d–f: *Hierarchical model as in Fig. 56.* d: *for black values of q the orbit of* $1 - q$ *goes neither to* 1 *nor to* $\infty$. ($-1 \leq Re\ q \leq 2.65$, $-2 \leq Im\ q \leq 2$). e: *detail from* f ($1.92 \leq Re\ q \leq 1.97$, $0.88 \leq Im\ q \leq 0.93$); *for black values of q the orbit of* 0 *does not go to* 1. f: *for black values of q the orbit of* 0 *does not go to either* 1 *or to* $\infty$. ($-.8 \leq Re\ q \leq 3.5$, $-2 \leq Im\ q \leq 2$)

Mandelbrot set reappears so completely intact, without a single bud missing. We admit that the universality expressed in these observations strikes us with awe even though the mathematicians A. Douady and J. H. Hubbard have recently succeeded in developping a theoretical understanding for this computer experimental result.

# 9 Yang-Lee Zeros

For those readers familiar with the formalism of canonical statistics, we present here some facts about the location of the Yang-Lee zeros for the Ising model of magnetism.

Let every lattice site $i$ carry a spin $s_i$ which can take on the values $s_i = +1$ and $s_i = -1$. The spins interact with their nearest neighbors (indicated by $<i, j>$) and with an external magnetic field $H$. The total energy of a configuration $\{s_i\}$ is given by

$$(9.1) \quad E\{s_i\} = -J \sum_{\langle ij \rangle} s_i s_j - H \sum_i s_i.$$

For $J > 0$ this describes ferromagnetic interaction, for $J < 0$ antiferromagnetic interaction. The partition function is the sum of the Boltzmann weight factors for all possible configurations:

$$(9.2) \quad Z_N(T, H) = \sum_{\{s_i\}} \exp(-E\{s_i\}/k_B T) = \sum_{\{s_i\}} \exp\left(K \sum_{\langle ij \rangle} s_i s_j + L \sum_i s_i\right),$$

where the quantities $K = J/k_B T$ and $L = H/k_B T$ are introduced as convenient temperature and magnetic field variables, respectively. The free energy per lattice site is obtained from $Z_N$ in two steps:

$$(9.3) \quad f_N = -\frac{k_B T}{N} \ln Z_N; \qquad f(T, H) = \lim_{N \to \infty} f_N.$$

The first step involves taking the logarithm. This produces singularities at the zeros of $Z_N$ which makes them candidates for representing the magnetic phase transition. But note that for real temperatures $T$ and fields $H$ the exponentials in $Z_N$ are all positive so there can be no zeros. This prompted Yang and Lee to look into complex values of $T$ and $H$. They argued that the second step in (9.3) might be all important. In taking the thermodynamic limit $N \to \infty$, more and more complex zeros might be generated and come closer and closer to the real axis. Even though there is no real zero for any given $N$, accumulation points of zeros may develop on the real axis as $N \to \infty$, and thereby generate the observed singularities of f(T, H) at the physical phase transitions. This expectation is corroborated in the following examples.

### *H Dependence for an Arbitrary Lattice*

For a fixed temperature $T$ (that is, fixed $K$) the partition function $Z$, as a function of $H$ (that is, of $L$), has a particularly simple form. If we use $z = \exp(-2L)$ as a convenient complex field variable, we obtain (9.2) in the form

$$(9.4) \quad Z_N = z^{-N/2} \sum_{n=0}^{N} p_n(K) z^n,$$

where $n$ is the number of spins with $s_i = -1$. Up to the trivial factor $z^{-N/2}$, $Z_N$ is a *polynomial* in $z$ with positive coefficients $p_n(K)$.
The theorem of Lee and Yang [LY] states that for ferromagnetic interaction, $K>0$, this polynomial has all its zeros on the unit circle in the $z$-plane for every $N$. For real $T$, this means that all singularities of the free energy lie on the imaginary $H$-axis. If the zeros approach the real $H$-axis in the limit $N \to \infty$ (which they do for $T < T_c$), they do so at $H=0$. This remarkable result is independent of the lattice structure and of the spatial dimension of the model. The interactions between nearest neighbors need not even be isotropic.

### *One-Dimensional Lattice for H=0*

Here we consider a chain of $N$ spins without an external magnetic field. In order to also study the influence of the boundary conditions on the behavior of the zeros, we look at two cases. First, let both ends of the chain be free, and, second, we identify site $N+1$ with 1, i.e., we close the chain to a ring.
With $x = \exp(2K)$ we obtain the partition functions

free chain:

$$(9.5) \quad Z_N^f = 2x^{-(N-1)/2}(x+1)^{N-1},$$

ring:

$$(9.6) \quad Z_N^r = x^{-N/2}[(x+1)^N + (x-1)^N].$$

It turns out that the thermodynamic behavior (for $N \to \infty$) in the physical temperature region $x>0$ is the same in both cases, even though the distributions of the zeros are drastically different. $Z_N^f$ has one zero of order $(N-1)$ at $x=-1$, while $Z_N^r$ has its $N$ zeros on the imaginary $x$-axis at

$$(9.7) \quad x_k = i \cot \frac{(2k+1)\pi}{2N}, \; k=0, 1, \ldots, N-1.$$

In the parametrization $x = is$, the density of these zeros in the limit $N \to \infty$ is

$$(9.8) \quad \mu(s) = \frac{1}{N}\frac{dk}{ds} = \frac{1}{\pi}\frac{1}{1+s^2}.$$

These results confirm the general rule that there is no phase transition at positive temperatures in one-dimensional magnets. When (9.7) is expressed in the complex temperature $T$ by means of $x = \exp(2J/k_B T)$, it indicates a phase transition-like singularity at $T=0$.

### *Two Dimensional Lattice for H=0*

One of the great achievements in theoretical physics was the exact solution of the two dimensional Ising problem by the Norwegian chemist L. Onsager [On]. He showed for the first time in a specific example that the canonical formalism can produce singularities. The partition function $Z_N$ for a quadratic lattice with different interaction $(J_1, J_2)$ in each of the two directions can be written in closed form [Kau]:

$$Z_N = \prod_{r=1}^{m} \prod_{s=1}^{n/2} \left( \cosh 2K_1 \cosh 2K_2 - \cos \frac{2\pi r}{m} \sinh 2K_1 - \cos \frac{2\pi s}{n} \sinh 2K_2 \right), \tag{9.9}$$

where $K_i = J_i/k_B T$ and $N = m \cdot n$. The zeros of the partition function are the zeros of the factors in (9.9). For the isotropic case $K_1 = K_2 = K$, M.E. Fisher noted in 1964 that these zeros are all on the unit circle in the sinh $(2K)$-plane [Fi]. The anisotropic case has only been treated recently [SK, SC]. The result: the zeros can be distributed over an *area* in the plane. That area has a conic shape near the real axis so that the general expectation of Yang and Lee is still fulfilled: a single real accumulation point of zeros indicates the location of the physical phase transition.

# 10 Renormalization

The idea of renormalization can be understood as a successive thinning out of the degrees of freedom in the partition function. The $N$-particle problem is transformed into an $N'$-particle problem with $N' < N$, whereby the temperature $T$ and the magnetic field $H$ may also have to be renormalized. Assume in the following that there is no external magnetic field, $H=0$. Then the precise formulation of the renormalization idea is this: find a transformation

$$(10.1) \quad T \mapsto T' = R(T)$$

such that the $N$-particle partition function can be expressed by the simpler $N'$-particle partition function as

$$(10.2) \quad Z_N(T) = Z_{N'}(T')\varphi_N.$$

The factor $\varphi_N$ accounts for the different levels of the energy zero in the $N$ and $N'$ particle systems and is not important in the further discussion.
By means of (9.3), the relation (10.2) can be rewritten as a functional equation for the free energy per particle:

$$(10.3) \quad f(T) = \frac{N' \cdot T}{N \cdot T'} f(R(T)) + g(T),$$

where g is usually nonsingular at the interesting critical point $T = T_c$. The critical point is a repulsive fixed point of the renormalization transformation, $T_c = R(T_c)$. By (10.2) or (10.3), the leading singularity of $f(T)$ at $T_c$ is closely related to the linear part of R in $T_c$. This is the basis on which Wilson derived the critical power laws. We want to use these recursion relations to find the Yang-Lee zeros. For that purpose we shall discuss three examples where (10.2) can be derived rigorously. In all these cases $Z_N$ is a polynomial of degree $d(N)$ in a suitably defined temperature variable x. Consider $Z_N$ and $Z_{N'}$ as given in terms of their zeros $x_l$ $(l=1, \ldots, d(N))$ and $x'_m$ $(m=1, \ldots, d(N'))$ respectively. (10.2) can then be written

$$(10.4) \quad \prod_{l=1}^{d(N)} (x - x_l) = \varphi_N \prod_{m=1}^{d(N')} (x' - x'_m).$$

Forget $\varphi_N$ for the sake of simplicity in the argument (it will be taken into account in the specific examples). When the left hand side of (10.4) is zero, the right hand side must also be zero. Therefore the zero sets of $Z_N$ and $Z_{N'}$ must be related by the renormalization transformation $x \mapsto x'$. This transformation is usually not one-to-one. Since $N' < N$, $Z_{N'}$ is an easier object to study than $Z_N$. In order to find the zeros of $Z_N$, we may first determine the zeros of $Z_{N'}$ and then take their preimages with respect to $R$ ($x$ is a preimage of $x'$ provided $R(x) = x'$).
This reasoning may be iterated from $N$ to $N'$ to $N''$ etc. until eventually we deal with a trivial 2-particle partition function $Z_2$. Iterating the renormalization

backwards from the zeros of $Z_2$ we find all the zeros for the cases of more and more particles until finally we arrive at the thermodynamic limit.
When the renormalization transformation $R$ happens to be a rational map as it does in the examples below, one can conclude from the work of Julia and Fatou that backward iteration leads typically towards the Julia set of $R$. In fact, if $R$ has degree $D$ (represent $R$ as the quotient of relatively prime polynomials; then the degree of $R$ is the maximum of the degrees of these polynomials), then typically (except for special situations) a point $x$ has $D$ preimages with respect to $R$. The above statement relating the Yang-Lee zeros and the Julia set is intuitively obvious if one assumes that the zeros of $Z_2$, the initial points for the backward iteration, are in a basin of attraction of some fixed point of $R$. Then taking successive preimages naturally leads to the boundary of that basin in the thermodynamic limit, and according to (2.6) in Special Section 2 this is the Julia set. Moreover every point of the Julia set is approximated in this way because inverse orbits on the Julia set are dense (see (2.5)). This establishes the identification of the Yang-Lee zeros in the thermodynamic limit with the Julia set of the renormalization transformation. Therefore, to discuss phase transitions it suffices to characterize the intersection of the Julia set with the real temperature axis. This remarkable connection was first observed by Derrida, DeSeze, and Itzykson [DDI].
The above intuitive argument is in fact supported by the following fact: Given any rational mapping $R$ with Julia set $J_R$. Then for any $\bar{x} \in \overline{\mathbb{C}} \setminus E$ one has that

$$J_R \subset \text{closure}\,\{x \in \overline{\mathbb{C}}: R^n(x) = \bar{x} \text{ for some } n\}. \tag{10.5}$$

The exceptional set $E$ is usually very small. Here is an example: let $P$ be a polynomial. Then $P(\infty) = \infty$ and $\infty \in E$. In fact, we see that the above set of preimages is just $\{\infty\}$, i.e. the approximation of $J_P$ through this set fails.

### *The One-Dimensional Ising Model*

The renormalization can be carried out quite easily for the linear chain by, say, summing over every other spin; this gives $N \mapsto N' = N/2$. In the notation used for the Ising model in Section 9, we have in the absence of a magnetic field:

$$\begin{aligned}
Z_N(x) &= \sum_{s_1 s_3 \dots} \; \sum_{s_2 s_4 \dots} x^{\frac{1}{2}(s_1 s_2 + s_2 s_3) + \frac{1}{2}(s_3 s_4 + s_4 s_5) + \dots} \\
&= \sum_{s_1 s_3 \dots} \left(x^{\frac{1}{2}(s_1+s_3)} + x^{-\frac{1}{2}(s_1+s_3)}\right) \left(x^{\frac{1}{2}(s_3+s_5)} + x^{-\frac{1}{2}(s_3+s_5)}\right) \dots \\
&= \sum_{s_1 s_3 \dots} \left(\sqrt{2\left(x+\frac{1}{x}\right)}\,(x')^{\frac{1}{2} s_1 s_3}\right) \left(\sqrt{2\left(x+\frac{1}{x}\right)}\,(x')^{\frac{1}{2} s_3 s_5}\right) \dots \\
&= Z_{N/2}(x')\varphi_N
\end{aligned}$$

with $\varphi_N = [2(x+1/x)]^{N/4}$ and the renormalization transformation

$$x' = R(x) = \frac{1}{2}\left(x + \frac{1}{x}\right). \tag{10.6}$$

This is a rational mapping of the complex $x$-plane onto itself which is very simple to characterize. Through a conjugation by means of $z=(x+1)/(x-1)$ we obtain the mapping

$$(10.7) \quad z' = \tilde{R}(z) = z^2.$$

The Julia set of this mapping is the unit circle in the $z$-plane, so that the Julia set of the process (10.6) is the imaginary axis in the $x$-plane (cf. Cayley's problem in Section 6).
We obtain the connection with the zeros of $Z_N(x)$ in the thermodynamic limit when we iterate the zeros of $Z_2(x)$ backwards. For the chain with free ends, this is

$$(10.8) \quad Z_2^f = \frac{2}{\sqrt{x}}(x+1),$$

which, by chance, has its zero $x_0 = -1$ in the superstable fixed point of the mapping $R$. In this special case, the iteration does not lead away from $x_0$, which matches with eq. (9.5). In contrast to this, if we take the closed ring, we get

$$(10.9) \quad Z_2^r = 2\left(x + \frac{1}{x}\right),$$

with zeros at $\pm i$. These are already in the Julia set of the mapping $R$, so that the inverse iteration in the limit $N \to \infty$ yields a set of zeros that is dense in the Julia set (being the imaginary axis in the complex $x$-plane).

### *The Standard Hierarchical Lattice with Potts Spins*

Let us consider the lattice with the recursive structure shown in Fig. 55.
At the $n$-th stage this has $4^{n-1}$ bonds which, at the $(n+1)$-th stage, are each replaced by two new lattice sites and four bonds. Generalizing the Ising spin somewhat, we assume that each lattice site carries a *Potts* spin which can have $q$ different states $s_i = 1, \ldots, q$ and which interacts with neighboring spins if and only if the two states are the same:

$$(10.10) \quad E\{s_i\} = \sum_{\langle ij \rangle} E_{ij}, \quad E_{ij} = \begin{cases} -J & \text{if } s_i = s_j \\ 0 & \text{else.} \end{cases}$$

The partition function for the lattice at stage $n$,

$$(10.11) \quad Z_{(n)}(T) = \sum_{\{s\}} \exp\left(-E\{s_i\}/k_B T\right)$$

is a polynomial of degree $4^{n-1}$ in the variable $x = \exp(J/k_B T)$, with positive integer coefficients.
The reduction from the $n$-th stage to the $(n-1)$-th stage can be achieved easily with the help of a summation over the states of the last $2 \cdot 4^{n-2}$ spins incorporated. The result is [DDI, PR3]

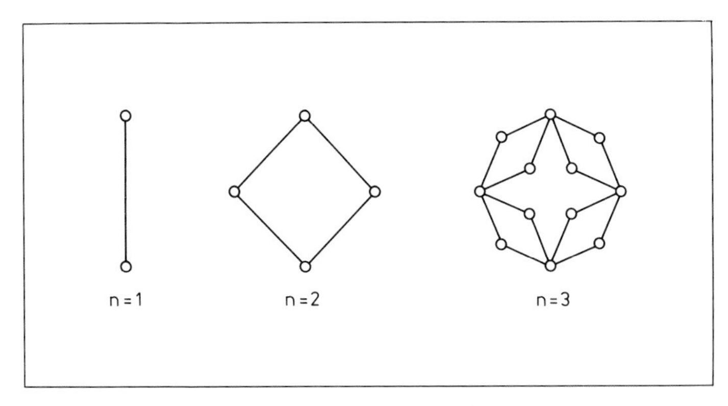


*Fig. 55. The standard hierarchical lattice.* ○ = *lattice sites;* — = *bonds*

(10.12) $Z_{(n)}(x) = Z_{(n-1)}(x')\varphi_{(n)}$

with

(10.13) $x' = R_q(x) = \left(\frac{x^2+q-1}{2x+q-2}\right)^2$

and

(10.14) $\varphi_{(n)} = (2x+q-2)^{2 \cdot 4^{n-2}}$

The renormalization transformation (10.13) is a rational mapping of degree 4, so that each zero of $Z_{(n-1)}$ corresponds to four zeros of $Z_{(n)}$ as pre-images with respect to $R_q$. If one starts with the one zero $x_0 = 1-q$ of

(10.15) $Z_{(1)} = q(x+q-1)$,

one obtains all zeros of $Z_{(n+1)}$ by $n$-fold iteration of the transformation $R_q^{-1}$. In the limit $n \to \infty$, this gets arbitrarily close to the Julia set of $R_q$ provided $1-q$ does not belong to the exceptional set E mentioned with (10.5).
Another way to see the connection of the Julia set of $R_q$ and the singularities of the free energy $f$ is to recall the functional equation (10.3), with $T$ replaced by x. Since $f$ is singular at the repulsive fixed point $x_c$ of $R_q$, it must also be singular at $x \in R_q^{-1}(x_c)$. Iterating this argument and using (2.5) we see that the Julia set of $R_q$ is a set of singularities of $f$.
To characterize the mapping $R_q$ and its Julia sets, we note that it has two superattractive fixed points $x=1$ and $x=\infty$. These correspond to the temperatures $T=\infty$ and $T=0$ ($J>0$) and thus to the cases of complete disorder and complete order, respectively. The nature of the boundary between the domains of attraction of these two points, i.e. the phase boundary between para- and ferromagnetism, depends on the fate of the critical points of $R_q$. These are the 6 points

(10.16) $\{1, \infty, 1-q, \pm\sqrt{1-q}, (2-q)/2\}$.

The points 1 and $\infty$ have already been recognized as attractive fixed points. $(2-q)/2$ is mapped to $\infty$ at the first step and stays there. $\pm\sqrt{1-q}$ both go to 0, so that we only need to follow the orbits of $1-q$ and 0. It can be shown that these orbits are in some sense complementary [PR3, 4], so that in fact only the point $x=0$ has to be investigated. There are the following three possibilities:

*(i)* $x=0$ belongs to the ferromagnetic phase, that is, $R_q^n(0) \to \infty$ with $n \to \infty$.
*(ii)* $x=0$ belongs to the paramagnetic phase, that is, $R_q^n(0) \to 1$ with $n \to \infty$.
*(iii)* $x=0$ is not attracted to either 1 or infinity.

According to a theorem of Fatou (see Section 3), there can be no further attractors in the first two cases, whereas case *(iii)* allows for one or two more phases (as long as $x=0$ does not belong to the Julia set).

### *Another Hierarchical Lattice*

Figure 56 shows the structure of another hierarchical lattice.
The motivation for the study of this model is that in the special case of $q=2$, which corresponds to Ising spins, it leads us to expect an antiferromagnetic phase. The renormalization transformation turns out to be a rational mapping of degree 6,

$$(10.17) \quad x' = S_q(x) = \left( \frac{x^3 + 3(q-1)x + (q-1)(q-2)}{3x^2 + 3(q-2)x + q^2 - 3q + 3} \right)^2 .$$

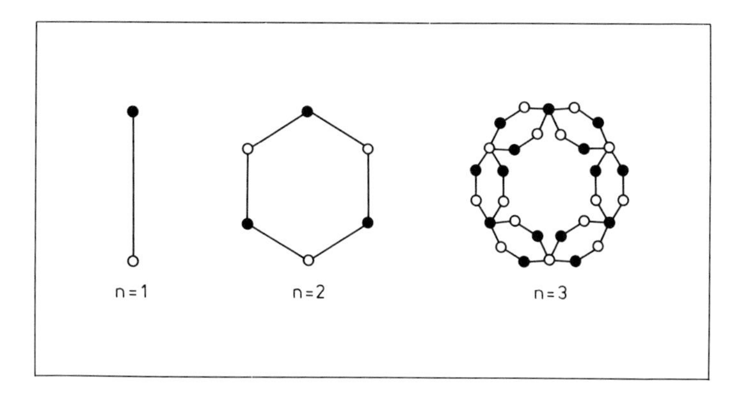


*Fig. 56. A hierarchical lattice for antiferromagnetic coupling*

The zeros of the partition function $Z_{(n)}(x)$ are again the pre-images of $x_0 = 1-q$ under iteration of $S_q^{-1}$. The Yang-Lee phase boundary is again the Julia set of $S_q$. $x=1$ and $x=\infty$ are, as above, superstable attractors independently of $q$, and the analysis of the critical points shows that here, too, only the fate of the points 0 and $1-q$ is relevant. In contrast to the standard model, the two points do not give us equivalent results (cf. Fig. 54 d and f) so that the number of alternatives increases and the over-all picture becomes more complicated.

# REFERENCES


*Books*

[AS] Abraham RH, Shaw C (1985) Dynamics, The Geometry of Behavior, I, II, III. Aerial Press, Santa Cruz

[Co] Cole KC (1985) Sympathetic Vibrations. Morrow W, New York

[DH] Davis, PJ, Hersh R (1981) The Mathematical Experience. Birkhäuser, Boston

[De1] Devaney RL (1986) Introduction to Chaotic Dynamical Systems. Benjamin-Cummings, Menlo Park

[EW] Eigen M, Winkler-Oswatitsch R (1975) Das Spiel. Piper, München

[Fal] Falconer KJ (1985) The Geometry of Fractal Sets. Cambridge University Press, Cambridge

[H] Haken H (1981) Erfolgsgeheimnisse der Natur. Deutsche Verlagsanstalt, Stuttgart

[Ma1] Mandelbrot BB (1982) The Fractal Geometry of Nature. Freeman, San Francisco

[My1] May RM (1974) Model Ecosystems. Princeton University Press, Princeton

[MS] Moser J, Siegel CL (1971) Lectures on Celestial Mechanics. Springer-Verlag, Grundlehren Bd 187

[PR1] Peitgen H-O, Richter PH (1984) Harmonie in Chaos und Kosmos. Bremen

[PR2] Peitgen H-O, Richter PH (1984) Morphologie komplexer Grenzen. Bremen

[Sch] Schuster HG (1984) Deterministic Chaos – An Introduction. Physik Verlag, Weinheim

[Sm] Smith CS (1980) From Art to Science. MIT Press, Cambridge, Mass

[St] Stanley HE (1971) Introduction to Phase Transitions and Critical Phenomena. Oxford University Press

*Survey Articles*

[Be] Berry MV (1978) Regular and Irregular Motion. In: Jorna S (ed) Topics in Nonlinear Dynamics. Amer Inst of Physics Conf Proceedings 46: 16–120

[Bl] Blanchard P (1984) Complex Analytic Dynamics on the Riemann Sphere. Bull Amer Math Soc 11: 85–141

[Bro] Brolin H (1965) Invariant Sets Under Iteration of Rational Functions. Arkiv f Mat 6: 103–144

[DT] Deker U, Thomas H (1983) Unberechenbares Spiel der Natur – Die Chaos-Theorie. Bild der Wiss 1: 63–75

[Dou] Douady A (1985) L'étude Dynamique des Polynômes Quadratiques Complexes et ses réinvestissements. Soc Math de France, Assemblée Générale

[G1] Großmann S (1969) Analytic Properties of Thermodynamic Functions and Phase Transitions. In: Madelung O (ed) Festkörperprobleme IX. Vieweg 207–245

[G2] Großmann S (1984) Discrete Nonlinear Dynamics. In: Velarde MG (ed) Non-Equilibrium Cooperative Phenomena in Physics and Related Fields. ASI-Series, Plenum Press, New York

[G3] Großmann S (1983) Chaos-Unordnung und Ordnung in Nichtlinearen Systemen. Phys Blätter 39: 139–145

[Ho] Hofstadter DR (1982) Strange Attractors: Mathematical Patterns Delicately Poised Between Order and Chaos. Scientific American 245/5, 16–29, Spektrum der Wiss, January 7–17

[My2] May RM (1976) Simple Mathematical Models with Very Complicated Dynamics. Nature 261: 459–467

[NF] Nelson DR, Fisher ME (1975) Soluble Renormalization Groups and Scaling Fields for Low-Dimensional Ising Systems. Annals of Phys 91: 226–274

[PR3] Peitgen H-O, Richter PH (1986) Fraktale und die Theorie der Phasenübergänge, Phys. Blätter 42: 9–22

[PSH] Peitgen H-O, Saupe D, Haeseler F v (1984) Cayley's Problem and Julia Sets. Math Intelligencer 6: 11–20

[RP] Richter PH, Peitgen H-O (1985) Morphology of Complex Boundaries. Berichte d Bunsenges f Physikal Chemie 89: 571–588

[Ru1] Ruelle D (1980) Strange Attractors. Math Intelligencer 2: 126–137

*Selected Original Work*

[Ba] Barna B (1956) Über die Divergenzpunkte des Newtonschen Verfahrens zur Bestimmung von Wurzeln algebraischer Gleichungen, II. Publ Mathematicae Debrecen 4: 384–397

[Br] de Branges L (1985) Proof of the Bieberbach conjecture. Acta Math 154: 137–152

[Cam] Camacho C (1978) On the local structure of conformal mappings and holomorphic vector fields in $c^2$. Soc Math France, Astérisque 59–60: 83–93

[Ca] Cayley A (1879) The Newton-Fourier imaginary problem. Amer J Math II 97

[CGS] Curry J, Garnett L, Sullivan D (1983) On the iteration of rational functions: Computer experiments with Newton's method. Commun Math Phys 91: 267–277

[DDI] Derrida B, De Seze L, Itzykson C (1983) Fractal Structure of Zeroes in Hierarchical Models. J Statistical Physics 33: 559–569

[De2] Devaney RL (1984) Julia sets and bifurcation diagrams for exponential maps. Bull Amer Math Soc 11: 167–171

[DH1] Douady A, Hubbard JH (1982) Iteration des polynomes quadratiques complexes. CRAS Paris 294: 123–126

[DH2] Douady A, Hubbard JH (1984) On the dynamics of polynomial-like mappings. preprint

[DH3] Douady A, Hubbard JH (1984) Etude dynamique des polynomes complexes. Publications Mathematiques D'Orsay

[Fa] Fatou P (1919/1920) Sur les équations fonctionelles. Bull Soc Math Fr 47: 161–271, 48: 33–94, 208–314

[Fei] Feigenbaum M (1978) Quantitative Universality for a Class of Nonlinear Transformations. J Statistical Physics 19: 25–52

[Fi] Fisher ME (1965) The Nature of Critical Points, Lectures in Theor Phys VII c, 1–160, University of Colorado Press, Boulder

[GT] Großmann S, Thomae S (1977) Invariant Distributions and Stationary Correlation Functions of One-Dimensional Discrete Processes. Zeitschr f Naturforschg 32a: 1353–1363

[Ha] Haeseler F v (1985) Über sofortige Attraktionsgebiete superattraktiver Zyklen. Dissertation Universität Bremen

[Ju] Julia G (1918) Sur l'iteration des fonctions rationnelles. Journal de Math Pure et Appl 8: 47–245

[Kad] Kadanoff LP (1966) Scaling laws for Ising Models near $T_c$. Physics 2: 263–272

[Kau] Kaufman B (1949) Crystal Statistics. II. Partition Function Evaluated by Spinor Analysis. Phys Review 76: 1232–1243

[Lo] Lorenz EN (1964) The problem of deducing the climate from the governing equations. Tellus XVI: 1–11

[LY] Lee TD, Yang CN (1952) Statistical Theory of Equations of State and Phase Transitions. II. Lattice Gas and Ising Model. Phys Review 87: 410–418

[Lj] Ljubich MJ (1983) Entropy properties of rational endomorphisms of the Riemann sphere. Ergod Th & Dynam Sys 3: 351–385

[Ma2] Mandelbrot B (1980) Fractal aspects of the iteration of $z \to \lambda(1-z)$ for complex $\lambda$, $z$. Annals NY Acad Sciences 357: 249–259

[MSS] Mane R, Sad P, Sullivan D (1983) On the dynamics of rational maps. Ann Scient Ec Norm Sup 16: 193–217

[On] Onsager L (1944) Crystal Statistics. I. Two-Dimensional Model with an Order-Disorder Transition. Phys Rev 65: 117–149

[PPR] Peitgen H-O, Prüfer M, Richter PH (1985) Phase Transitions and Julia Sets. In: Ebeling W, Peschel M (eds) Lotka-Volterra-approach to cooperation and competition in dynamic system. Akademie-Verlag, Berlin

[PR4] Peitgen H-O, Richter PH (1985) The Mandelbrot Set in a Model for Phase Transitions. In: Hirzebruch F et al (eds) Proc 25, Bonner Mathem Arbeitstagung 1984. Springer Lecture Notes in Math 1111: 111–134

[PPS] Peitgen H-O, Prüfer M, Schmitt K (1985) Global Aspects of the Continuous and Discrete Newton Method: A Case Study. preprint, University of Utah

[Ru2] Ruelle D (1982) Repellers for real analytic maps. Ergod Th & Dynam Sys 2: 99–108

[SK] Saarlos W v, Kurtze DA (1984) Location of Zeros in the Complex Temperature Plane: Absence of Lee-Young Theorem. Journal of Physics A 17: 1301–1311

[Si] Siegel CL (1942) Iteration of Analytic Functions. Ann of Math 43: 607–612

[SC] Stephenson J, Couzens R (1984) Partition Function Zeros for the Two-Dimensional Ising Model. Physica 129 A: 201–210

[Su] Sullivan D (1982–1983) Quasi conformal homeomorphisms and dynamics I. II. III. preprint. IHES

[Wi] Wilson KG (1971) Renormalization Group and Critical Phenomena I. Phys Rev B4: 3174–3183, II. Phys Rev B4: 3184–3205

[YL] Yang CN, Lee TD (1952) Statistical Theory of Equations of State and Phase Transitions. I. Theory of Condensation. Phys Rev 87: 404–409

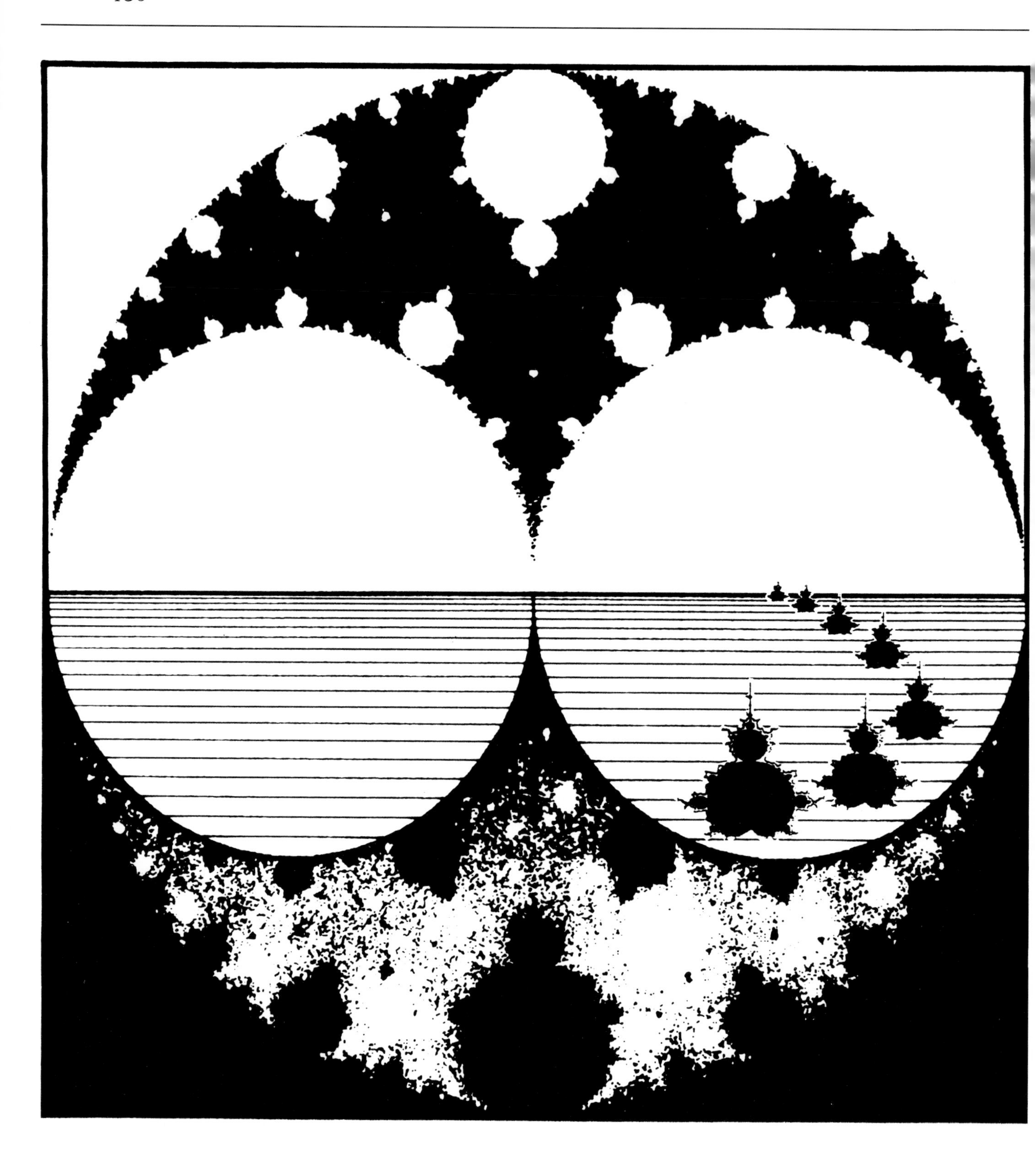

# INVITED CONTRIBUTIONS

## Fractals and the Rebirth of Iteration Theory

BENOIT B. MANDELBROT

No more than six years ago! Only ten and twenty-odd years ago!
On many days, I find it hard to believe that only six years have passed since I first saw and described the structure of the beautiful set which is celebrated in the present book, and to which I am honored and delighted that my name should be attached. No more than twenty-odd years have passed since I became convinced that my varied forays into unfashionable and lonely corners of the Unknown were not separate enterprises. No one had seen any unity between them, other than provided by my personality; yet, around 1964, they showed promise of consolidating one day into an organized field, which I proceeded to investigate systematically. And no more than ten years have passed since my field had consolidated enough to justify writing a book about it, hence giving it a name, which led me to coin the word *fractal geometry.*
The beauty of many fractals is the more extraordinary for its having been wholly unexpected: they were meant to be mathematical diagrams drawn to make a scholarly point, and one might have expected them to be dull and dry. It is true that the poet wrote that Euclid gazed at beauty bare, but the full and continuing appreciation of the beauty of Euclid demands hard and long training, and perhaps also a special gift. To the contrary, it seems that nobody is indifferent to fractals. In fact, many view their first encounter with fractal geometry as a totally new experience from the viewpoints of aesthetics as well as science. From these viewpoints, fractals are indeed as new as can be.
From the purely mathematical viewpoint, the situation is more complex and very interesting. Many scientific theories start their life by borrowing wholesale from already organized areas of mathematics, but in this instance, no such organized area existed. It is the fractal geometry built for the needs of science that has had the unexpected effect of uniting several old and noble – but narrow – streams of mathematics into a single active one, while also reawakening several other streams that had been dormant.
The custom is to start historical accounts with the more distant past, and then to move forward to the present. But I would like to reverse the order in the present case. Before my memory falters, let me retell the very beginnings of the beautiful set featured in this book. Some of the computer drawings reproduced here are the very earliest illustrations of the shapes they represent. Today, I view them as valuable antiques. Yesterday, they seemed out-of-date and pitifully primitive. And in 1980, when they lit my life with intellectual and aesthetic revelations, they were the best one could do at Harvard University, where I was a Visiting Professor of Mathematics in 1979–80. The basement of the Science Center housed its first Vax computer (brand new and not yet "broken in"); to view the pictures, we used a Tektronix cathode ray tube (worn out and very faint), and our hard copies were printed on a Versatec device no one knew how to set up properly. The home base from which I had come to visit was the Thomas J. Watson Research Center of IBM, at Yorktown

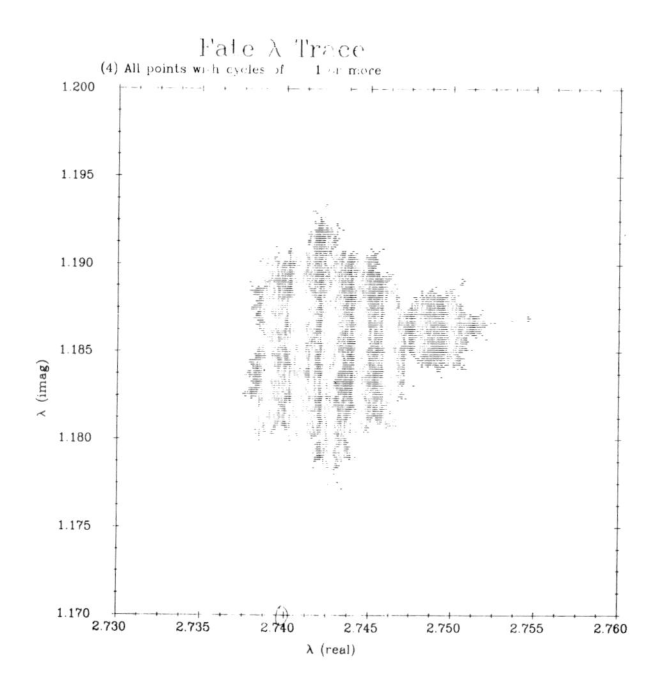


*Fig. 57.*
*First detailed picture of an "island molecule" in the Mandelbrot set for $z \mapsto z^2 - c$ (1st March 1980)*

Heights, NY, where I had found academic freedom as an IBM Fellow. But these 1980 pictures may perhaps dispel any impression that I lived and prospered scientifically by dipping into IBM's roomfulls of up-to-date custom equipment.
Adding to the irony, the beautiful graphics produced at IBM for my earlier 1977 Essay had come from an unstable set-up, of hardware that had once been scrapped and of special software that never came to be documented.
My excellent programmer in 1980 was Peter Moldave, a course assistant "moonlighting" as an unpaid research assistant.
It is necessary at this point to back up to the year 1978–79, when my excellent assistant at IBM was Mark R. Laff. I had become entranced with fractals that are invariant under *non linear* transformations, whereas the fractals I had first studied were self-similar, that is, invariant under linear transformation. Why this new infatuation? The reason is that I had read an obituary of Henri Poincaré, written by Jacques Hadamard, and this had drawn my attention to several abandoned branches of mathematics that promised to involve interesting fractals whose actual structure was for all purposes unknown.
First, we played with a construct originally considered by Poincaré himself in the 1880's. It is called the "limit set of a Kleinian group". More precisely, we played with the following related problem: Given several circles in the plane, describe the structure of the set of points that is invariant – unchanged – under ordinary inversions in either of these circles. Alternatively, starting with an arbitrary initial point, one subjects it to an infinite sequence of inversions with respect to the given circles, and one seeks the shape to which these strings of inversions "attract" the initial point. To my pleasure and surprise, my skills in experimental mathematics made me discover an explicit construction. After

the fact, it is almost obvious, yet it had eluded the pure mathematicians since the 1880's.

Then we went on to mindless fun, drawing many examples of the shapes known as "Julia sets". These sets enter in what is called "the theory of iteration of rational maps of the complex plane". This theory was dormant in 1979, having reached its high point long before, around 1918, with famous papers by G.Julia and P.Fatou. Why return to these papers? I had read or scanned them at age twenty, under the prodding of an uncle (a prominent pure mathematician, and specialist in complex analysis), and they had been incredibly influential in my life. Their immediate effect in 1945 was to help me forsake the pattern one usually follows in the study of mathematics. The fact that Julia was one of my teachers at Polytechnique led to no change of mind. But thirtyfive years later I took a leading part in the revival of the theory of iteration, which had the belated effect of bringing me closer to the mainstream of mathematics than I would have thought possible. We accumulated beautiful drawings of Julia sets by the bushel (one Julia set had been drawn independently by John H.Hubbard, who showed it to us). It was nice to understand intuitively, at long last, what Julia and Fatou had really been after. And in addition, nearly all Julia sets proved to be extraordinarily beautiful.

However, the fun soon wore off for me, and I picked a serious task: I selected a family of rational maps with one complex parameter, and set out to investigate the domains of this parameter such that the dynamics of the map converge to stable limit cycles of various sizes. Let this set be called $M'$. Somehow, I felt that in order to achieve a set $M'$ having a rich structure, it was best to pick a complicated map (every beginner I have since then watched operate has taken the same tack). To make sure that there could be a parameter value leading to certified chaos, I picked the map $z \mapsto c(1+z^2)^2/z(z^2-1)$. For $c=1/4$, one has a map singled out by Samuel Lattes, whose dynamics is known to be chaotic. Then I made $c$ vary over the whole complex plane. There being no criterion for the existence of a stable limit cycle for this map, we let the computer grind on with an exploratory algorithm, and we ended with a highly structured but very fuzzy "shadow" of $M'$. A recent rendering is shown in Fig. 58.

In 1979 all we could see was a very blotchy version. It sufficed to show that the topic was worth pursuing, but had better be pursued in an easier context.

This is why I went back to the quadratic map $z \mapsto z^2 - c$. Since it always has a stable fixed point at infinity, the interesting problem is that of classifying the fixed cycles that are bounded. The quadratic map is the simplest of all maps, and the only one for which everything depends upon a single parameter.

In the early 1970's, the investigations of P.J. Myrberg, concerning the case where $c$ is real, had become widely known and were being extended in various ways. But – astonishing as it may sound – no one was tackling their complex plane extension. I felt that the known properties of the real quadratic maps would provide a running test for the case when $c$ is complex. To be able to move on rapidly, I felt it was safe to take an exploratory short cut – whose full mathematical justification was beyond my analytic powers, and as a matter of fact still happens to be lacking. Fatou had shown in 1906 that for certain

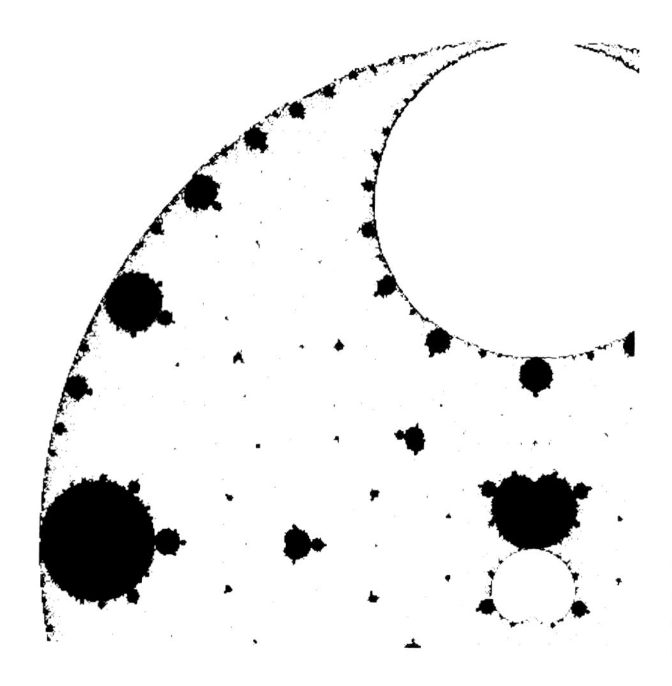

*Fig. 58. One quarter of the Mandelbrot set for the map $z \mapsto (z^2+1)^2/z(z^2-1)$ as redrawn in 1982*

c's the point at infinity attracts all the complex plane, except for a Julia set that is very "thin" and forms what is now called a "fractal dust". A dust bounds no domain of any sort, hence cannot bound a domain of attraction. Therefore, the set $M'$ I was seeking had to be a subset of the set of c's for which the Julia set is not a dust, hence is connected.

The latter set, denoted by $M$, is the set to which my name has become attached. The reason I picked it is that Julia has given a straightforward criterion that is particularly easy to program for the quadratic map: $c$ belongs to the set $M$ if and only if the point $z_0=0$ (called "critical point") *fails* to converge to infinity. A first test used the quadratic map in the alternative form $z \mapsto \lambda z(1-z)$. After few iteration steps on a rough grid, we saw that the set $M$ includes the very crude outline of the two discs $|\lambda|<1$ and $|\lambda-2|<1$. Two lines of algebra confirmed that these discs were to be expected here, and that the method was working. We also saw, on the real line to the right and left of the above discs, the crude outlines of round blobs which I call "atoms" today. They appeared to be bisected by intervals known in the Myrberg theory, which encouraged us to go on to increasingly bold computations. For a while, every investment in computation yielded increasingly sharply focussed pictures. Helped by imagination, I saw the atoms fall into a hierarchy, each carrying smaller atoms attached to it. We verified that the points where the big disc-shaped atoms carry smaller atoms are as expected. Thus, I saw geometric implementations, not only for the familiar Myrberg sequence of successive binary bifurcations, but for every sequence of bifurations of arbitrary order.

After that, however, our luck seemed to break; our pictures, instead of becoming increasingly sharp, seemed to become increasingly messy. Was this the fault of the faltering Tektronix? To make sure, I arranged a brief visit to Yorktown. When our Harvard program was run on the IBM mainframe, we ob-

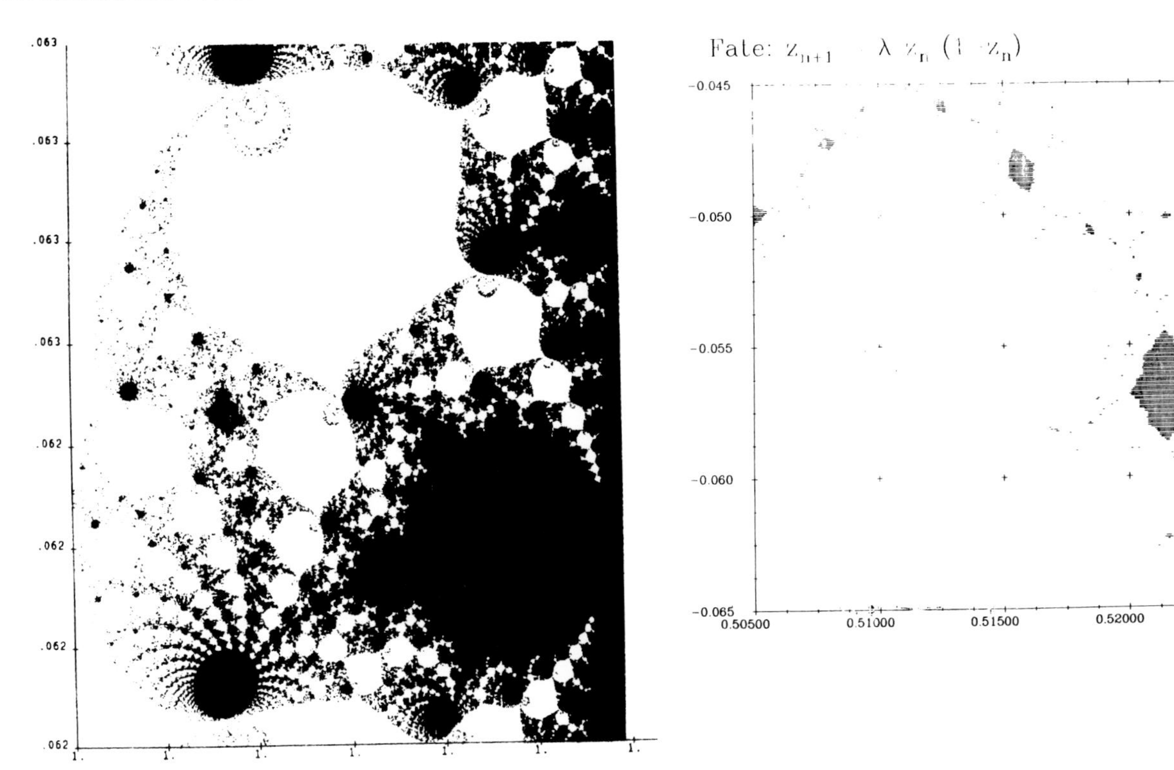


*Fig. 59. First picture (June 1980) of a detail of the Mandelbrot set at a bifurcation of order 100*

*Fig. 60. Julia set for a parameter value that lies in one of the "island molecules" of the Mandelbrot set (April 1980)*

tained an illustration already published on the top of Plate 189 of my 1982 Essay. The mess had failed to vanish! In fact, as you can check, it showed signs of being systematic. We promptly took a much closer look. Many specks of dirt duly vanished after we zoomed in. But some specks failed to vanish; in fact, they proved to resolve into complex structures endowed with "sprouts" very similar to those of the whole set M. Peter Moldave and I could not contain our excitement. Some reason made us redo the whole computation using the equivalent map $z \mapsto z^2 - c$, and here the main continent of the set $M$ proved to be shaped like each of the islands! Next, we focussed on the sprouts corresponding to different orders of bifurcation, and we compared the corresponding off-shore islands. They proved to lie on the intersections of stellate patterns and of logarithmic spirals! Figure 59 is an example drawn at IBM in the summer of 1980, using a mainframe and Tektronix hard copy; this example corresponds to 100-fold bifurcation.
Combined with these blowups of the set $M$, we were running pictures of the Julia sets for values of $c$ that lie within the island molecules. What we saw appeared to split into many islands, each of these a reduced scale version of the Julia set corresponding to a matching value of $c$ in the continental molecule of the set $M$ (Fig. 60). However, Julia's criterion implied that this appearence had to be misleading. While island interiors cannot overlap, the gap between islands had to be partly spanned by a smaller island, and so on ad infinitum.

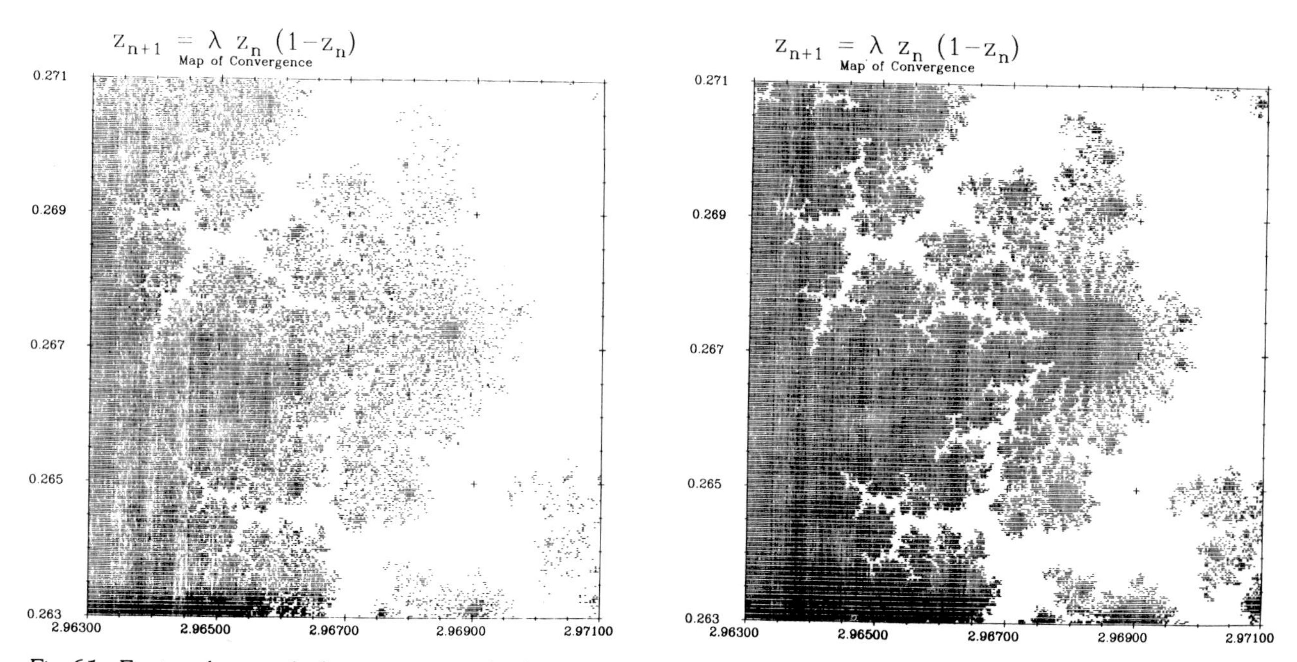


*Fig. 61. First and second of two successively sharper pictures of a detail of the Mandelbrot set (April 1980)*

Ultimately, the islands had to connect by their coastlines, adding up to a "devil's" polymer, whose "strands" were invisible because actual computation is necessarily limited to a lattice.

We continued to flip in this fashion between the set *M* and selected Julia sets *J*, and made an exciting discovery. I saw that the set *M* goes beyond being a numerical record of numbers of points in limit cycles. It also has uncanny "hieroglyphical" character: including within itself a whole deformed collection of reduced-size versions of all the Julia sets.

In terms of the set *M*, one special feature of the Myrberg theory, as developed by 1980, became very important. It implies that the real axis pierces a string of islands of the set *M*, and connects them by their coastlines, in the devil's fashion I have already described as being found in certain Julia sets. This suggested that the stellate structures we had noticed outside of the real axis are also due to the whole set *M* being a connected devil polymer.

When I faced this issue, however, I acted with excessive caution, which was wholly out of character, luckily proved temporary, and may have been brought up by a year sitting among pure mathematicians. Let me elaborate.

The varied and often conflicting influences I keep welcoming during my checkered career keep bringing layer upon layer of "sediment", which often affect me in peculiar ways. Ordinarily, I am insensitive to being accused by mathematicians of advancing insufficiently rigorous arguments, but on this occasion I allowed the mathematical component of my scientific persona to dominate. Though I had failed in 1980 to prove that my set *M* is connected, I should have stated its being connected as an experimental observation. But I did not have the nerve.

At that time, I was writing a paper that eventually appeared late in 1980 and

became very well-known. Part of this paper reports on a talk I had given in New York in December 1979, and (as is often the case) another part concerns newer developments. But, instead of discussing the set $M$ in the form I had been studying it, I defined in this paper an awkward surrogate, whose properties could be described in mathematically more "firm" fashion.
The connectedness of the set $M$ was thereby presented as a question to be answered, rather than a conjecture to be verified. Not until my 1982 book did I snap back into proper assertiveness. Soon, the issue became moot, when A. Douady and J. H. Hubbard gave a proof of the connectedness of the set $M$, and went on to study it in admirable detail.
While I have specifically been asked to use this contribution to tell things as I remember them happening, I may be writing in terms that are more self-revealing than my hosts had expected. But allow me to continue in the same vein when tackling the notion of fractal, because its definition had involved a different example of the same uneasy – though thrilling – coexistence of the varied threads of a scientific persona. In 1975, I coined the term *fractal* in order to be able to give a title to my first Essay on this topic. But I stopped short of giving a mathematical definition, because I felt this notion – like a good wine – demanded a bit of aging before being "bottled". The shapes I was investigating and called fractals in my mind all share the property of being "rough but self-similar". The word similar does not always have the pedantic sense of "linearly expanded or reduced", but it always conforms to the convenient loose sense of "alike". This loose sense was needed, for example, to include the Poincaré and Julia sets discussed above.
"Rough and self-similar" was a "middle third" possibility for systematic geometric discourse, attempting to squeeze itself between the two possibilities to which this discourse had been limited until then. The first possibility was exemplified by the whole body of Euclid, which is concerned with shapes that are extremely ordered and smooth. (The detail of a curve in Euclid is also self-similar, but in an uninteresting fashion; all curves are locally straight, and the straight line is self-similar.) The second old possibility was concerned with shapes of arbitrary complication and roughness. Today, these shapes deserve to be called "geometrically chaotic", but at that time I denoted them by the weaker Latin equivalent, "erratic". The thrust of my new field of investigation was to split chaos. One part was to continue to be left alone because it remained unmanageable, but a second part, less general yet very substantial, deserved to be singled out. To study it was important because of the innumerable occurences of self-similarity which I kept finding in nature, and it was feasible, precisely because of self-similarity.
However, the contrast between "orderly" and "disorderly chaos" was not yet a central issue when my 1975 Essay was being written. At that stage, it was still necessary to gain acceptance for chaos, by drawing a contrast between smooth and unsmooth shapes. I was pressed to draw this contrast firmly, by defining fractals in formal fashion. It happens that, long ago, chance had made me know of a concept called Hausdorff dimension, and I developed an intuitive understanding of it. Everyone else deemed it to be anything but intuitive; in fact, it was a very obscure notion for the great majority among working mathematicians, though a classical one for a few mathematicians who prac-

ticed it. Had I failed to develop this knowledge and this intuition, fractal geometry would not be, I think.
But, in due time, I came to recognize that, in fact, my intuition had always dealt with different forms of a broader concept of what I call "fractal dimension". The strength of Hausdorff's definition of fractal dimension is to distinguish between the categories of "smooth" and "chaotic". But its weakness is to fail to distinguish between the categories of "rough and self-similar" and "geometrically chaotic". This is so because the definition is very general, which is desirable in mathematics. But in science its generality was to prove excessive: not only awkward, but genuinely inappropriate.
Yet, this feature was not yet apparent in 1975, while my use of a fractional dimension in the context of my self-similar models was proving to be a shock to scientists. My response was not that of a pragmatic scientist, but that of a mathematician. I rushed to the protective embrace of an existing notion, and I declared in 1977 to be sets whose Hausdorff dimension is a fraction, or otherwise exceeds their topological dimension. This definition left out many "borderline fractals", yet it took care, more or less, of the frontier "against" Euclid. But the frontier "against" true geometric chaos was left wide open! I know that definitions matter little, but this one can still be improved upon.
Finally, this examination of history brings us to the distant roots of the events of twenty-odd years ago. By the most exacting standards of the philosopher and the historian, few thoughts are wholly new. When an endeavor is unimportant, its claims to novelty do not deserve to be researched. But important developments – and fractal geometry does show promise of being seen as one – demand careful probing. Even before it became established, I subjected its intellectual background to the most exacting standards, and fully reported the results in my books. On its contribution to science and aesthetics, the conclusion is that there was not even an inkling of fractal geometry before my work. In fact, the near-total absence of unrecognized forerunners is very surprising. Historical search did reveal a few obscure quotes (by J. Perrin, H. Steinhaus and few others) that point out to chaos as something to think about, but there was no follow-up of these thoughts.
On the other hand, my book quotes innumerable famous mathematicians who worked in the period 1875–1925, including Poincaré, Cantor, Peano, Hausdorff, Sierpinski. Do I think, therefore, that fractal geometry was "invented" a hundred years ago? Not at all! My reasons in quoting these authors invariably combine strong praise with strong blame. The praise was for their having invented certain constructions that eventually I was able to link together and found to be invaluable. The blame was for their having failed to see and develop a kinship among their constructions, and for having handled each of them as a "monster" or an "exceptional" set, which thoroughly missed their true significance. This historical context helps explain why, as I have said, my fractal geometry came as a total surprise to everyone – and most of all to the practitioners of the mathematical discipline called "real analysis"

◁ *Fig. 62. "Signature" of Benoit B. Mandelbrot (executed by Eriko Hironaka)* which was born of the same constructions of circa 1900. I am very flattered that these surprising ideas came very soon to be viewed as "natural" and "unavoidable".

I should stop here. Much more about the history of fractals as I recall it has been printed recently, and these details have no direct bearing on the gorgeous book for which – before I was carried away – I was writing this essay.

*References*

Mandelbrot BB (1975) Les objets fractals, Flammarion, Paris

Mandelbrot BB (1977) Fractals: Form, Chance, and Dimension, Freeman, San Francisco

Mandelbrot BB (1980) Fractal aspects of the iteration of $z \rightarrow \lambda z(1-z)$ for complex $\lambda$ and $z$. In: Nonlinear Dynamics, Helleman RHG (ed). Annals New York Acad. Sciences 357, pp 249–259

Mandelbrot BB (1982) The Fractal Geometry of Nature, Freeman, San Francisco

In his own words: Benoit Mandelbrot (1985) An Interview by Barcellos A, in: Albers, DJ and Alexanderson GL (eds) Mathematical People, Birkhäuser, Boston, pp 205–225

# Julia Sets and the Mandelbrot Set

ADRIEN DOUADY

Quadratic Julia sets, and the Mandelbrot set, arise in a mathematical situation which is extremely simple, namely from sequences of complex numbers defined inductively by the relation

$$z_{n+1} = z_n^2 + c,$$

where $c$ is a complex constant. I must say that, in 1980, whenever I told my friends that I was just starting with J.H. Hubbard a study of polynomials of degree 2 in one complex variable (and more specifically those of the form $z \mapsto z^2 + c$), they would all stare at me and ask: Do you expect to find anything new? It is, however, this simple family of polynomials which is responsible for producing these objects which are so complicated – not chaotic, but on the contrary, rigorously organized according to sophisticated combinatorial laws.

The behavior of the sequence mentioned above depends upon the following data: the parameter $c$ and the initial point $z_0$. Julia sets are defined by fixing $c$ and letting $z_0$ vary in the field of complex numbers, while the Mandelbrot set is obtained by fixing $z_0 = 0$ and varying the parameter $c$. If you take $z_0$ far from 0, then the sequence tends very quickly towards infinity. This is also true if for some $n$ the point $z_n$ is far from 0. To make a quantitative statement, if for some $n$ the modulus $|z_n|$ of $z_n$ (i.e. the distance from the point $z_n$ to the origin) is greater than $|c| + 2$, then the modulus of $z_{n+12}$ is greater than the ratio of the volume of the known universe (all the way to the most remote quasars) to the volume of a proton. But there are values of $z_0$ for which the sequence $(z_n)$ never goes far away but remains bounded. For a given $c$, these values form the *filled-in Julia set* $K_c$ of the polynomial $f_c: z \mapsto z^2 + c$. The actual *Julia set* consists only of the boundary points of $K_c$ (In this paper, I shall often refer to $K_c$ inaccurately as the Julia set). Of course it may happen that $K_c$ has no interior point, in which case it coincides with its boundary (in other words the Julia set coincides with the filled-in Julia set).

Obviously, the Julia set depends on the choice of the parameter $c$, but the surprise here is that it depends enormously on it, so that by varying $c$ you obtain an incredible variety of Julia sets: some are a fatty cloud, others are a skinny bush of brambles, some look like the sparks which float in the air after a firework has gone off. One has the shape of a rabbit, lots of them have sea-horse tails ...

There are two major classes of Julia sets: some are in one piece – we say they are connected – while the others are just a cloud of points, which we call a Cantor set. For a mathematician, this is a good opportunity to define a new set: The set of values of $c$ for which $K_c$ is connected. I called it the *Mandelbrot set M* because Benoît Mandelbrot was the first one to produce pictures of it, using a computer, and to start giving a description of it. As I said above, it can also be defined as the set of values of $c$ for which, starting with $z_0 = 0$, the sequence $(z_n)$ remains bounded (the equivalence of these two definitions is a consequence of a theorem proved in 1919, independently, by Fatou and Julia).

When you look at the Mandelbrot set, the first thing you see is a region limited by a cardioïd, with a cusp at the point .25 and its round top at the point −.75. Then there is a disk centered at the point −1 with radius .25, tangent to the cardioïd. Then you see an infinity of smaller disk-like components, tangent to the cardioïd, most of which are very small. Attached to each of those components, there is again an infinity of smaller disk-like components, and on each of these there is attached an infinity of smaller disk-like components, and so on. But that is not all! If you start from the big cardioïd, go to the disk on the left, again to the component on the left, and keep going, you will tend to a point called the Myrberg-Feigenbaum point, situated at −1.401. . . . Now, the segment from this point to the point −2 is contained in *M*. And on this segment, there is a small cardioïd-like component, with its cusp at −1.75 (its center is at −1.754877666. . .), accompanied by its family of disk-like satellites just like the big cardioïd. Actually there are infinitely many such cardioïd-like components. There are also cardioïd-like components off of the real axis. B. Mandelbrot discovered one centered at $-0.1565201668+1.032247109\,i$, and many others. In fact he showed that there are an infinite number of them. They are so tiny that it is hard to distinguish them from stains on the computer picture (except for the fact that they arise symmetrically). However, if you make an enlarged picture, you will discover for each of them the cardioïd shape and its company of disk-like components. And this is not all . . . All of these cardioïd-like components are linked to the main cardioïd by filaments, charged with small cardioïd-like components, each of which is accompanied by its family of satellites. These filaments are branched according to a very sophisticated pattern, of which detailed combinatorial studies have been made. Because of these filaments, the set *M* is itself connected. The proof of this fact is by no means obvious, and there are still lots of open questions. For instance, one cannot prove up to now that the "filaments" I mentioned are actually arcs of curves that one can parametrize continuously.

Julia sets are among the most beautiful fractals. Most of them are self-similar: if you look at the boundary of a given $K_c$ with a microscope, what you see does not depend essentially upon where you look, nor on the magnifying power of the microscope. In contrast, the Mandelbrot set does *not* possess this property of self-similarity: certainly *M* contains an infinite number of small copies of itself, so that wherever you look in the boundary of *M* with a microscope, you will see some small copies of *M*. But these copies of *M* are embedded in a network of filaments whose aspect depends very much on where you are looking. Moreover, if you see two copies of comparable size, the ratio of their distance to their size depends highly not only on the place where you are looking, but also on the magnifying power of the microscope.

To understand how the shape of the Julia set corresponding to a given point $c$ in *M* is related to the location of $c$ in *M*, is perhaps the main purpose of the study of the Mandelbrot set.

When one wants to study a bounded set *K* of complex numbers, one way is to compute the *potential* that it creates, and the *external arguments* of points on its boundary. The definition of these comes from electrostatics. Imagine a capacitor made of an aluminium bar shaped in such a way that its cross-section is $K$, placed along the axis of a hollow metal cylinder. Set the bar at potential 0 and the cylinder at a high potential. This creates an electric field in the region between the cylinder and the bar. An electric potential function is also established in this region. Assume that the radius of the cylinder is large (with respect to the chosen unit), that its height is large compared to its radius and that the length of the bar is equal to this height. We restrict our attention to the plane perpendicular to the axis of the cylinder, through its middle. In this plane, the electric potential defines *equipotential lines* enclosing the set $K$ (which is the cross section of the bar), and following the electric field you get field-lines, which I call the *external rays* of $K$. Each external ray starts at a point $x$ on the boundary of $K$, and reaches a point $y$ of the great circle which is the cross section of the cylinder (practically at infinity). The position of $y$ is identified by an angle, which we call the *external argument* of $x$ with respect to $K$. If there are several accesses to $x$ from outside of $K$ (for instance if $K$ is tree-like and $x$ is a branch point of $K$), then there is one external ray in each access, and the point $x$ has several external arguments. Giving a list of all external arguments for each point that has several is actually a good way of providing information about the shape of $K$: it tells you how many branches there are at each branch point, and gives you an idea of their size.

Well, this is a beautiful program, but it is usually very hard to carry out, even for simple shapes, such as polygons. But here comes a fantastic surprise: For the Julia sets, as well as for the Mandelbrot set, the potential and the external arguments are very easy to compute. The potential is essentially given by the *escape time*, which is defined as follows: take a large radius $R$, say $R=100$. Then if you consider a Julia set $K_c$, the escape time for a point $z$ outside of $K_c$ is the first $n$ for which $z_n$ (defined by $z_0=z$ and $z_{n+1}=z_n^2+c$) has modulus greater than $R$. For the Mandelbrot set $M$, you define the escape time using the same sequence, but with $z_0=0$, so that it depends only on $c$.

If $N$ is the escape time, the potential is approximately $(\text{Log } R)/2^N$, and you can make corrections to get its value with great accuracy. This works for the Mandelbrot set as well as for the Julia sets. Most of the pictures of Julia sets and of details of the Mandelbrot set that you see are obtained by computing the escape time for each point of the screen, and coloring the point accordingly using a color chart fed into the computer. H.-O. Peitgen always pays extreme attention to the choice of the color chart upon which the aesthetic effect depends greatly.

So, the potential produces the intricate and beautiful pictures which you can admire. However, for the mathematician, external arguments are an absolutely fascinating thing to compute. It is difficult to communicate this feeling without being technical, but I shall do my best. I begin with another surprise: most of the remarkable points, in Julia sets as well as in the Mandelbrot set, have external arguments which are rational (i.e. fractions with integer numerator and denominator), when you measure angles with the whole turn (360°) as a unit (Fig. 63). Moreover, rationals with even denominator and rationals with odd denominator play completely different roles. Why is this so? For Julia sets, it is not too hard to understand. Each Julia set $K_c$ is provided with its dy-

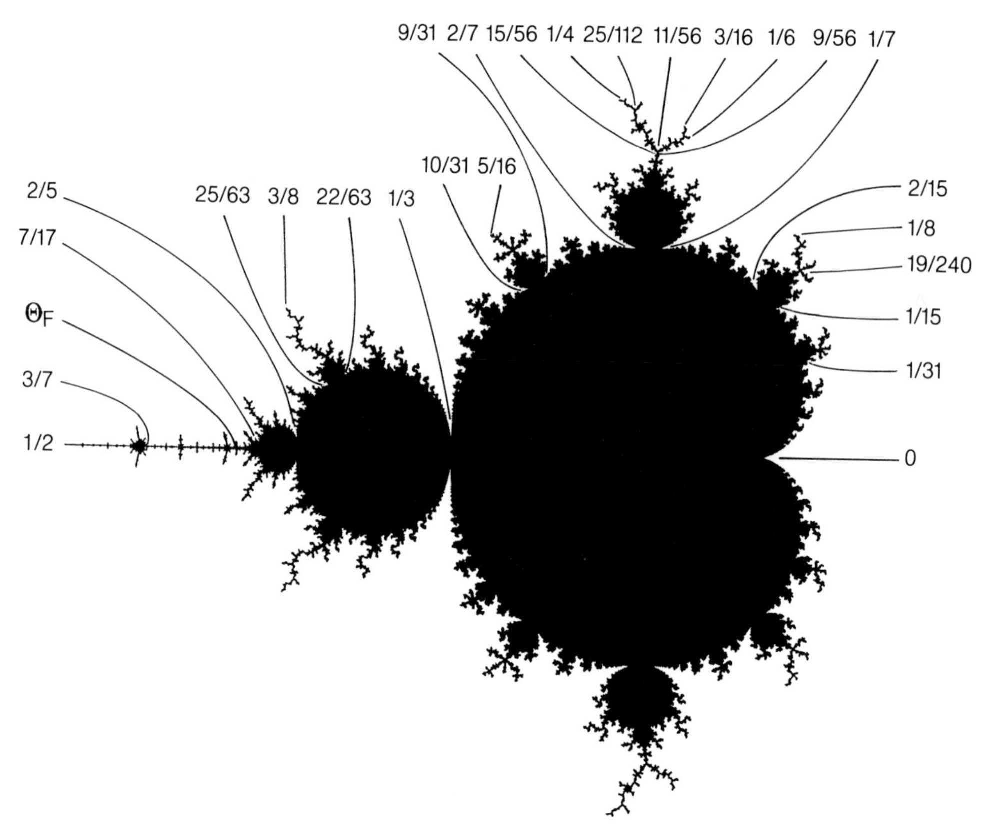


*Fig. 63. External arguments of points in M (the external argument $\Theta_F$ of the Feigenbaum-Myrberg point is not rational; it has been proved to be transcendental)*

namics, namely the map $f_c$: $z \mapsto z^2+c$ that maps $K_c$ onto itself. There are points which are periodic under $f_c$; this means that they belong to a cycle $(z_1, z_2, \ldots, z_k)$ with $f_c(z_1)=z_2$, $f_c(z_2)=z_3$, ..., $f_c(z_k)=z_1$. There are also points which are preperiodic: this means that they fall into a cycle after a finite number of steps. Now it turns out that the map $f_c$ has the effect of doubling the external arguments. So points which are periodic under $f_c$ have external arguments which are periodic under doubling, points which are preperiodic under $f_c$ will have external arguments which are preperiodic under doubling. But you have to remember that you are dealing with angles: a whole turn does not count, so that doubling means doubling in the ordinary sense for angles that are less than half a turn, while for angles between half a turn and a whole turn, you must double in the ordinary sense and subtract one turn. Angles which are periodic under this operation are just those which are measured by rational numbers with odd denominators, while those which are nonperiodic but preperiodic are precisely those which are rational with even denominators (this is a relatively easy exercise of arithmetic).

If a point is really remarkable, say a branch point or the end of a seahorse tail, you can expect its image to be remarkable too, and, since there are not so many highly remarkable points, it is reasonable to expect it to be periodic, or at least preperiodic. This essentially explains our statement in the case of Julia sets.

Let us come back now to the Mandelbrot set. As we have said, there is first a big cardioïd. For values of $c$ inside this cardioïd, the polynomial $f_c$ has an attracting fixed point. Then there is a disk centered at $-1$; for $c$ in this disk, $f_c$ has an attracting cycle of order 2. There are 3 components where $f_c$ has an attracting cycle of order 3, two attached to the cardioïd (the biggest ones except for the disk of period 2), and one on the real axis with a cusp at $-1.75$. There are 6 components of period 4, 15 of period 5, 27 of period 6, 63 of period 7, 120 of period 8, 252 of period 9, and so on. In each of these components, there is one value of $c$ (we call it the *center*) such that 0 is periodic for $f_c$. And each of them has on its boundary a point called the *root:* it is either a point by which it is attached to a bigger component, or a cusp in which a filament arrives.

Besides the centers and roots of the components, corresponding to the existence of an attracting cycle, there are other remarkable points in the Mandelbrot set, called the *Misiurewicz points.* They are the values of $c$ for which 0 is non-periodic, but preperiodic under $f_c$. They are usually branch points or ends of filaments. Now I can tell you about external arguments of points in $M$: Centers of components have no external argument, since they are inside. The root of each component has two external arguments, which are rational with odd denominator; each Misiurewicz point has one or several external arguments (one if it is the end of a filament, three or more if it is a branch point), which are rational with even denominator.

I have to describe how we compute external arguments, and then how we exploit this information once we have collected it. As soon as you have seen a Julia set and understood its dynamics (this component goes onto that one, this big one wraps twice onto this small one, and so on), it is very easy to compute the external angle of any point. The map $f_c$ has two fixed points; I call $\alpha$ the one on the left and $\beta$ the one on the right. The point $\beta$ has external argument 0, the point $\alpha$ usually has several external arguments. The point $-\beta$, the point symmetric to $\beta$ with respect to 0, has external argument $1/2$. You can draw the *spine* of $K_c$: it is an arc which joins $-\beta$ to $\beta$ staying in $K_c$. Now choose a point $x$ in the boundary of $K_c$, and if there are several ways to reach $x$ from outside $K_c$ (remember there are 2 if $x$ is on a string, more if $x$ is a branch point), select one of them. There is one external ray landing on $K_c$ at $x$ for this choice, and here is how you compute its argument. First observe that, if $x$, or at least the selected access, is above the spine, the argument is clearly between 0 and $1/2$, while if it is below the spine, the argument is between $1/2$ and 1. Next, if say the argument is between 0 and $1/2$, you want to know whether it is between 0 and $1/4$ or between $1/4$ and $1/2$. Apply $f_c$, whose effect is to double the argument, and look where $x$ goes together with the selected access. If it goes above the spine, you know the double of the argument is between 0 and $1/2$, so the argument is between 0 and $1/4$, and if it goes below the argument is between $1/4$ and $1/2$. One more step and you will know in which interval of length $1/8$ the argument lies. Actually, following the orbit of $x$ together with its access, and marking 0 each time it comes above the spine and 1 each time it comes below, you obtain the digits after the decimal point of the expansion in base 2 of the argument. If $x$ is periodic, or just preperiodic, these digits become periodic, which is what you expect for a rational number (you can verify that fractions with odd denominators are just those numbers whose expansion in base 2 is periodic immediately after the decimal point).

Observe that, to make such computations, you do not need to know with precision the details of the wiggles of the Julia set. You just need to have some combinatorial information about the way its various parts are related to each other. J.H. Hubbard succeeded in concentrating this information in a diagram, that I call the Hubbard tree, and when I want to talk about a specific point in $M$, I find it much more convenient to draw its Hubbard tree than to give its coordinates.

Computing external arguments of points in the Mandelbrot set is not really hard, but why it works is much more mysterious, since $M$ carries no dynamics – it is just a control space. For Misiurewicz points, the computation is particularly simple: if $c$ is a Misiurewicz point, the external arguments of $c$ in $M$ are just the external arguments of $c$ in $K_c$ (the point $c$ belongs to $K_c$ because it is the image of 0). It is not possible to explain here why it is so, but we can relate it to a phenomenon which has been abundantly observed, and recently established by Tan Lei (a young Chinese student in France): if you look at $M$ with a microscope focused at $c$, what you see resembles very much what you see if you look at the Julia set $K_c$ with the same microscope still focused at $c$, and the resemblance tends to become perfect (except for a change of scale) when you increase the magnifying power. Since each Julia set is self-similar although Julia sets are very different from each other, it is no wonder that the Mandelbrot set is very diverse in its different regions.

For points which are roots of components of $M$, the situation is a little more complicated, but again, for such a point $c$, computing the external arguments of $c$ in $M$ reduces to computing the external argument in $K_c$ of some point $x$. It is not the point $c$, because in this case $c$ is inside $K_c$ and thus has no external argument, but some periodic point which somehow acts as a substitute for $c$. Anyhow the process is always the same: you plough in the dynamic plane (the plane of the Julia sets) and you harvest in the parameter plane (the plane of the Mandelbrot set).

Once you have computed lots of external arguments, in a given Julia set or in the Mandelbrot set, you can use this information to give a description of this set. For this, you must keep records of pairs of external arguments corresponding to the same point. Suppose for instance that, for some Julia set $K_c$, you know that the three external rays of argument 1/7, 2/7 and 4/7 land in the same point. Then you see rightaway that $K_c$ consists of 3 pieces connected at this point, one carrying all points having an argument between 1/7 and 2/7, one carrying points with an argument between 2/7 and 4/7, and the third one with the points having an argument between 4/7 and 1 or between 0 and 1/7. Moreover the point where these three pieces connect, which has 1/7, 2/7 and 4/7 as external arguments, is necessarily the fixed point $\alpha$ of the polynomial. Indeed, the image of the point with argument 1/7 is the point with argument 2/7; if these two points are the same it is a fixed point.

Collecting all possible information of this type, you can build a model of the set $K_c$ in the following way: You start from a disk D, its boundary marked with angles $t$. Each time you notice a pair $(t,t')$ with $t$ and $t'$ two arguments of *one*

point in $K_c$, you pinch D so as to make the corresponding points on the boundary of D coïncide. When there are three arguments for one point, you also collapse the triangle. At the end, the model you have constructed, the pinched disk, will hopefully have the same shape as the set $K_c$ you want to describe. There is a theorem by Caratheodory (a mathematician of the beginning of this century), which tells you when this procedure succeeds. The condition is a mathematical property of the set $K_c$ which is stated by saying it is *locally connected.* Cantor dusts are not locally connected, but most Julia sets are; however, we know some Julia sets which are not locally connected even though they are connected. There are also some which are probably locally connected, but for which we are not able to give a mathematical proof that they are. Such is the one represented in Map 25. For this particular Julia set, numerical experiments made by Manton, Nauenberg and Mike Widom tend to prove that it is locally connected, together with other interesting properties, but they do not provide mathematical certainty. (Incidentally, this specific Julia set is also an exception to a principle I have stated earlier: the origin, which is the center of symmetry, is surely a remarkable point; however, its external arguments, which we can compute, are not rational.)
This process of reconstructing a set out of the information we can collect about the external arguments of its points can be applied to the Mandelbrot set too. Using Hubbard trees, we can compute the external arguments of lots of points, then take a disk with its boundary graded from 0 to 1, and start pinching together 1/7 with 2/7, 10/63 with 17/63, the three points 9/56, 11/56 and 15/56, and so on. Continuing this process will literally chew some parts into filaments charged with droplets, and finally create a model for the Mandelbrot set (Fig. 64). This is what Thurston calls "the abstract Mandelbrot set". Now, does this model resemble the actual Mandelbrot set that you can admire pictured with many details here? For the mathematician, this is asking *is the Mandelbrot set locally connected?* We have a lot of numerical evidence that the answer to this question is yes, but up to the day when I am typing this sentence there is no mathematical proof. This is a very irritating situation, because we know the model pretty well, so that if we could answer this question we could say that we know essentially everything we want to know about the Mandelbrot set. Irritating ... or motivating; after all mathematicians live on problems more than on answers.
There is another open question concerning the Mandelbrot set, another question that puzzles all mathematicians working in this subject, and that no one could solve up to now. We know that any value of $c$ for which the polynomial $f_c: z \mapsto z^2 + c$ has an attracting cycle lies inside the Mandelbrot set, in other words, it belongs to $M$ and it is not a boundary point. The question is about the converse: *is any point which is interior to $M$ a value of $c$ such that $f_c$ has an attracting cycle?* To put it in another way, we have described components of $M$ (of the interior of $M$ to be more precise) which correspond to the presence of an attracting cycle in the z-plane. These components are well understood and classified: we know exactly how many there are for each order of the cycle, and so on. But, besides these, which for some obscure reason are called hyperbolic, shall we discover some day some *queer components,* corresponding to a completely different phenomenon than an attracting cycle? I

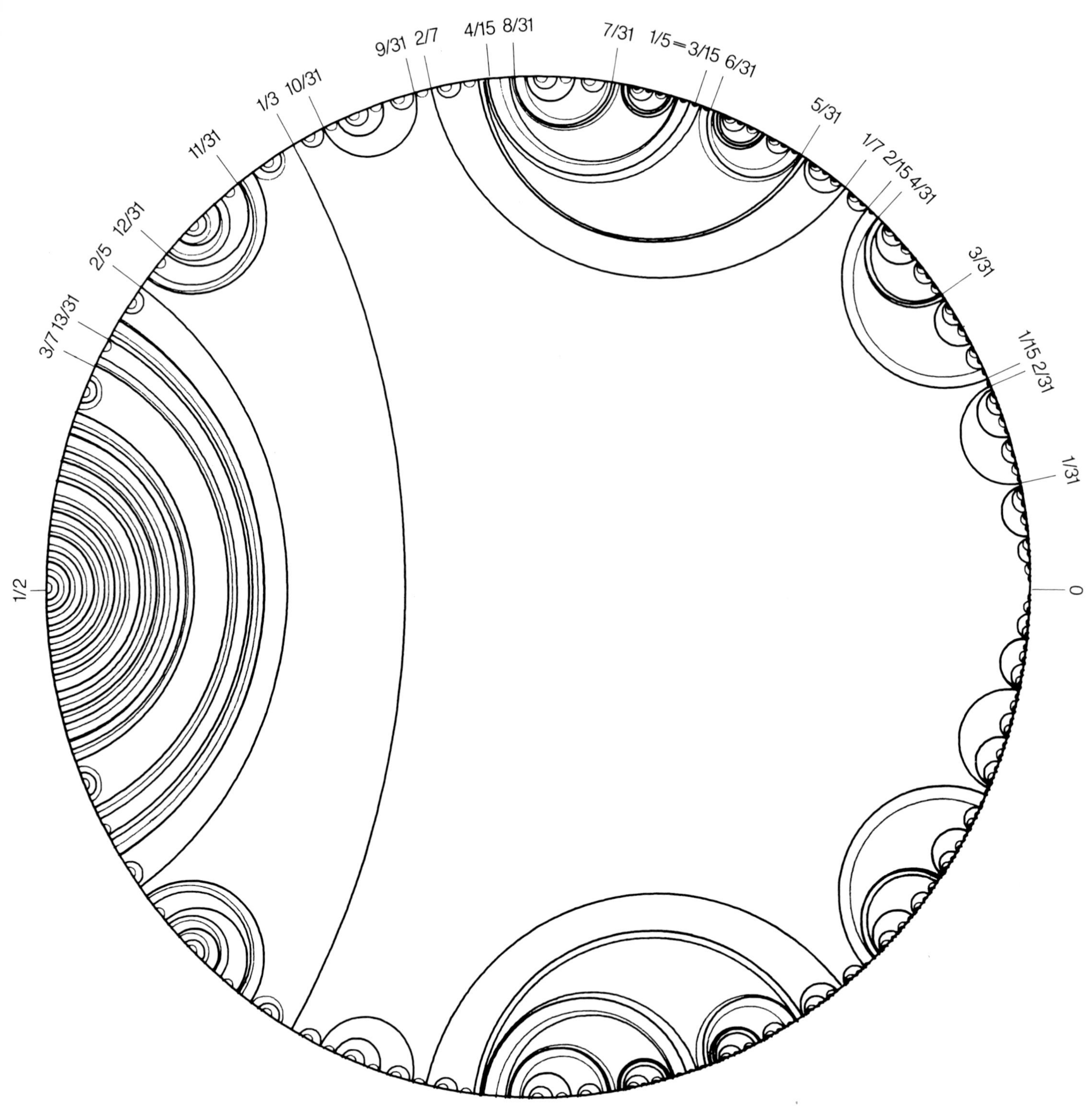


*Fig. 64. A model of the Mandelbrot set (Cesar Douady, 1982)*

am convinced that there is no queer component, nobody has ever seen one in the thousands of pictures of details of the Mandelbrot set that have been made, but again there is no mathematical proof. One could argue that computer pictures are difficult to interpret, and that they have always been analyzed by people who do not believe in queer components, so we cannot claim that their non-existence is established "beyond any reasonable doubt".

One can prove that the model of $M$ constructed by pinching a disk has no queer component. So that we are in the following curious situation: there are two conjectures that nobody has proved yet; the local connectivity of $M$ and the non-existence of queer components. But we can prove an implication between them. So, if tomorrow somebody proves that $M$ is locally connected, he or she will have proved at the same time that there is no queer component. This does not work the other way around: one of the two above conjectures is really stronger than the other one.

You may ask: Is it really worth taking all this trouble just for this very specific family of polynomials? Shall we have to start again from scratch when we have to iterate functions of another family? The answer is in a property that mathematicians call *structural stability* (physicists would rather speak of *universality*).

Let us start by a striking example. Map 2 shows a beautiful copy of the Mandelbrot set, on a background of green and orange zones, with brown tree-like features. However it comes from a problem which has nothing to do with iterating polynomials of degree 2. Namely, from the study of a rational function of degree 4 that arises in a special study of the nature of a magnetic phase transition. As Fig. 45 shows, similar structures are found in an analysis of Newton's method for solving polynomial equations of degree 3.

Newton's method is a way of improving an approximate solution of an equation. The idea, which goes back to Newton in the XVII[th] century, relies on a very simple computation. Suppose you know an approximate solution $x_0$ of an equation $f(x)=0$, close to an actual solution $x$ which you don't know. You can compute the small number $f(x_0)$ with accuracy, and also the value of the derivative $f'(x_0)$. Since $x$ is close to $x_0$, by definition of the derivative, $f'(x_0)$ is close to $\frac{f(x)-f(x_0)}{x-x_0}$, (which equals $-f(x_0)/(x-x_0)$ since $f(x)=0$). Therefore you get a good approximation of the desired correction $x-x_0$ by taking $-f(x_0)/f'(x_0)$, which you can compute with accuracy. So the improved approximate solution is the number $x_1=x_0-f(x_0)/f'(x_0)$. Iterating this procedure leads to a sequence of numbers $x_n$ which converges extremely quickly towards the actual solution. Roughly speaking the number of correct decimal digits is doubled at each step; all this works with complex numbers just as well.

So, if you start with a point close to a solution of the equation, you get a sequence which converges to this solution. Now what happens if you start with an arbitrary point $z_0$ in the plane of complex numbers and iterate Newton's method $z \mapsto z-f(z)/f'(z)$? This question was asked of Hubbard in 1977 by his students in their first year of university in Orsay. Thinking a few minutes, he proved that for equations of degree 2 the sequence always converges to the

root which is closest to $z_0$, unless $z_0$ is just on the bisector of the segment joining the two solutions, in which case the sequence $z_n$ remains on this line with a chaotic motion. "Now, he said, for equations of, say, degree 3, the situation seems more complicated, I will think of it and tell you next week". Probably the same dialogue occured in many places, because it is a very natural question. Actually, in the XIX[th] century, Cayley asked himself that very question, found the answer for equations of degree 2, and announced the case of polynomials of higher degree for a "forthcoming publication" which never appeared.
In between, computers had been invented, so that, if Hubbard and his students got nothing within a week, they got some experimental answers before the end of the semester. And that was the start of the involvement of Hubbard in "complex dynamics", and for me, by contamination.
The situation is actually pretty complicated. Some points are immediately attracted by one of the roots, some bounce here and there for a time, apparently at random, before they eventually land in the basin of immediate attraction of one root. You can prove that some points have a chaotic motion forever, but they are rare. Sometimes something else happens: *that a point is attracted by a cycle which does not correspond to a root* of the polynomial. Without the help of a computer, you can just list cases, but you are unable to discover any organization in the phenomena. But here there are questions which you can ask a computer. Choose an equation of degree 3, call $\alpha$, $\beta$ and $\gamma$ its solutions. Now the screen represents a window in the complex plane. You can ask the computer to color a pixel of the screen in *red* if the corresponding point is eventually attracted by $\alpha$, in blue if by $\beta$, in green if by $\gamma$, and in yellow if after iterating Newton's method, say, 1000 times, it does not reflect any attraction toward one of the roots. It takes some computational power to get a readable picture, but if you have good graphics, some of these pictures are marvelously beautiful. There are equations which are good – with respect to Newton's method –, for which almost all starting points lead to a root, so that you see no – or almost no – yellow. And there are the nasty equations for which Newton's method produces an attractive cycle, which leads to yellow patterns with interior points. Now here is an astonishing phenomenon: *these patterns reproduce Julia sets of polynomials of second degree!*
This situation also leads to copies of the Mandelbrot set: you may take a family of equations of third degree, given by a polynomial which depends on a parameter. Then, if you color, say, in black the values of the parameter which give good equations – good with respect to Newton's method – and in yellow the values which give nasty ones, you will see here and there small yellow copies of the Mandelbrot set. This is kind of magical, because there is not much relation between iterating a polynomial of degree 2 and searching for the roots of a polynomial of degree 3 by Newton's method! Of course, Newton's method is an iterative process, but what you iterate is the map

$$F_\lambda : z \mapsto z - \frac{f(z)}{f'(z)} = \frac{2a(\lambda)z^3 + b(\lambda)z^2 - d(\lambda)}{3a(\lambda)z^2 + 2b(\lambda)z + c(\lambda)},$$

an expression which is much more complicated than $z \mapsto z^2 + c$ from which the Mandelbrot set is defined.

I must be honest: it is not possible to ask for each pixel of the screen (representing now a value of the parameter) whether the corresponding equation is good or bad, by testing hundreds of thousand values of the initial point: on a pretty big computer this would take several minutes for each pixel, and more than a year for the whole screen! But it turns out that the barycenter of the roots, which you can compute easily (you can even choose your family of equations so that it is always 0), is a good test point: if this point goes to a root then the polynomial is good, and if after iterating a great number of times it does not come close to one of the roots, chances are good that the polynomial is bad. So the question you really ask the computer for each value of the parameter corresponding to a pixel of the screen, is: "does Newton's method converge to a root when you start at the test point?" When it does, you can ask to which root it converges, and color the pixel in red, blue or green according to the answer; you can also ask how long it takes before the point comes close to one of the roots, and make the color dark or light accordingly. This way you get a picture containing much information.
What happens is the following: in some area, the complicated function $F_\lambda$ that you iterate in Newton's method, or some finite iterate of it, resembles roughly, after proper rescaling, the simple function $f_c: z \mapsto z^2 + c$ (with some correspondence between the parameters $c$ and $\lambda$). The fact that 0 is trapped in some bounded region, which is characteristic of values of $c$ which are in the Mandelbrot set, corresponds to the test point of Newton's method being trapped in some region, preventing it from approaching the roots of the equation. Now, here is the miracle of structural stability: This rough resemblance between the $f_c$ and the $F_\lambda$ produces, for the sets defined in the variable plane (Julia sets) as well as for those defined in the parameter plane (Mandelbrot sets) a high similarity. This similarity does involve some rescaling, and also admits some (rather weak) distortion, but it preserves *all* of these sophisticated combinatorial features that one can detect using the computation of external arguments.
I can explain this in more general terms. The structural stability of the Mandelbrot set means that if you have a family of functions which resembles in some region the family of polynomials $(z \mapsto z^2 + c)$ and which is given by algebraic formulae, however complicated they may be (for those who know, the important condition is that it must be a holomorphic family), then the set produced by this family of functions has the same shape as the Mandelbrot set defined by the standard family. Let me stress one point: you would expect that each of the features of the Mandelbrot set has a domain of tolerance and that these domains are more restricted when you look at more subtle features, tending to zero when you really look in the finest details. But it is not so. It suffices that, after proper rescaling, the function you iterate differs from $z^2 + c$ by less than 1 for all values of $z$, and that $c$ does not exceed 4 in modulus, in order to insure that *all* the combinatorial features of the Mandelbrot set are preserved.
It is the structural stability of the Mandelbrot set which makes it really an important object to study. Problems of iteration arise in the study of the evolution of any system in any science from astronomy to biology or economics, so you may have to iterate any function. And the constant experience of mathematicians is that, when you have a problem with real numbers, it is always

enlightening to extend the functions you are considering to the domain of complex numbers. But the cases in which what you have to iterate is exactly a polynomial of second degree are scarce. However the cases in which it, or some of its finite iterates, looks roughly like a polynomial of second degree is, you may say, the general case. And this fact, together with structural stability, explains the following phenomenon, often observed: just take any dynamical system which can be modeled by the iteration of a function in the complex domain depending on a parameter, and classify the values of the parameter according to any dynamical property of the system you can think of. Then there is a good chance that you will see small copies of the Mandelbrot set in the parameter plane.

Actually this holds for the Mandelbrot set itself: A high iterate of a polynomial of degree 2 – which is a polynomial of degree $2^n$, thus very high – sometimes looks like a polynomial of degree 2 when you restrict it to a suitable region. It is this fact which is responsible for the presence of small copies of $M$ in $M$.

I would like to come back to Map 2 for a short remark. In this picture, you see green tongues of various sizes which touch the copy of the Mandelbrot set, each one at a different point. In the original Mandelbrot set, these points are exactly the points whose external arguments are fractions with a power of 2 in the denominator (halves, quarters or eighths for the great tongues, higher powers of 2 for smaller tongues). The other points in the boundary of the copy of $M$ can be reached by finding your way through the ramified structure that lies between the tongues, turning left or right at each branching. Now, if you keep track of the turns you make, marking 0 for a left turn and 1 for a right turn, you just obtain the binary expansion of the external argument of the point you have reached (more exactly of the corresponding point in the actual $M$). All this seems to indicate that external arguments are not just a mathematician's trick, a useful artifact, but that they really occur "in nature".

I would like to conclude with a reflexion concerning quantities of information. You may look at one of these pictures and, while you are fascinated by its internal rhythm, ask: what is the informational content of such a picture? If you mean by this: "how long should be a text which describes it?", then clearly the content is enormous. But if you think of the program which produces these pictures, then these programs are very short. The mathematical part takes 3 or 4 lines, most of the program deals with the framing and the color chart. We can think of the iteration process defined by the formula $z_{n+1} = z_n^2 + c$ as an extraordinarily efficient way to *develop the information* contained in the data (the value of $c$ for a Julia set, and the window for a detail of the Mandelbrot set), acting as a *key*. This phenomenon of developing information is also striking in biology: a transcription of all the genetic DNA of a human being (or any vertebrate) would take a hundred pages or so. Compare this with a treatise of anatomy, to which you should add one of endocrinology and one on innate behavior! Imagine scientists faced with the collections of Julia sets without knowing where they come from; would they not do just what zoologists did in the XIX$^{\text{th}}$ century: define phyla, classes, orders and genera, give a description of the specific features attached to each term of the classification, and so on?

Let us be clear: I am not claiming that Julia sets can provide a model for any biological phenomenon, but they are a striking example of how a very simple dynamical system can develop the small information contained in a key, and produce various highly organized structures.

# Freedom, Science, and Aesthetics

GERT EILENBERGER

This is a most unusual occasion that I have been asked to comment on. It is rather unusual for natural scientists to endeavor with such tenacity to bring their results and insights to the general public, but the *form* they use here is even more unusual! Instead of giving an abstract presentation in so many dry words, they have chosen pictures with a direct, universal appeal – a combination of mathematics and art!

Though I cannot contribute anything directly to this specific area of research, I do find the content of this exhibition inspiring, and I would like to add a few personal philosophical speculations concerning the possible significance these works may have for the physicist's understanding of the universe.

The occupation of natural science can be compared to the construction of a monumental building, say the cathedral of Cologne. We scientists are working on the cathedral of the scientific view of the universe. Though this cathedral, like the Cologne cathedral, has its practical uses, our reason for working on it is really, as it was expressed in the middle ages, for the glory of God. Only with *this* goal in mind does it indeed become a cathedral instead of a factory. And just as the workers in the construction-shack of the medieval cathedral are anonymous today, for it was the edifice itself that mattered not they, the contributions of most scientists will also remain anonymous. The cathedral is a communal work, and the scientists are the journeymen of a huge construction team, or, considering the worldwide extension of their activity, they are brothers in a worldwide order where the individual ought to retreat behind the great common work.

There is, though, an essential difference between scientific endeavors and the construction of a real cathedral: in science there are no blue-prints! Great surprises happen now and again, and the pictures presented here are extraordinarily surprising for the physicist, if perhaps not for the mathematician.

These pictures are not the result of trivial playing with the computer, amusing, but with no deeper meaning. The mathematical and physical insights on which these pictures are based are, in my estimation, the most exciting development since the discovery of quantum mechanics 60 years ago, insights which will again revolutionize our scientific view of the universe. Our cathedral will be completely transformed; it will lose its gothic coolness and take on baroque features!

The old picture of the universe, the scientist's creed, was formulated by the French mathematician and astronomer Laplace over 200 years ago. It can be paraphrased as:

*"If we can imagine a conciousness great enough to know the exact locations and velocities of all the objects in the universe at the present instant, as well as all forces, then there could be no secrets from this consciousness. It could calculate anything about the past or future from the laws of cause and effect."*

In such a deterministic world there would be no freedom and no chance. The deeds of a bank robber or the works of an artist would be predetermined. Scientists never really accepted this somewhat Calvinistic sounding predetermination for their everyday lives. But as far as they thought scientifically they couldn't escape it, since it was precisely this determinism that told them everything observable is, at least in principle, explainable scientifically, and *that* is an axiom that no scientist will give up easily! Even the great revolutions in physics during the first decades of this century, with which the names Planck, Einstein, and Heisenberg are connected, only moved this conflict to another mathematical level without finally having resolved it.

The researchers, however, were quite liberal with their interpretation of Laplace's credo – as is usual with credos. The most carefully constructed experiment is, after all, never *completely* isolated from the influences of the surrounding world, and the state of the system is never *exactly* known at any point in time. The absolute, mathematical precision that Laplace presupposed is not physically realisable – minute imprecision is always there, in principle.

What scientists then actually believed was this: from *approximately* the same causes follow *approximately* the same effects, in nature as well as in any good experimental set-up. And indeed this is *often* the case, especially for short time spans; otherwise we couldn't ascertain any natural laws or build any functioning machines.

But this apparently very plausible assumption is *not* universally true, and what's more, it does not do justice to the *typical* course of natural processes over long periods of time! This is, in a few short words, the exciting break through that has come from the investigations of so-called dynamical systems.

The art works exhibited here can be interpreted as the pictorial visualisations of such dynamical systems; these are motions, the time-change of mathematical quantities which can correspond to measurable quantities. These changes take place according to clear rules which are like the laws of nature.

The fine structures in these pictures are manifestations of the fact that the tiniest deviations at the beginning of a motion can lead to huge differences at later times – in other words, miniscule causes can produce enormous effects after a certain time interval. Of course we know from everyday life that this is *occasionally* the case; the investigation of dynamical systems has shown us that this is *typical* of natural processes.

Now, what does this say about freedom, i.e. about the possibility, in principle, of free decision, on which we base our idea of ourselves as responsibly acting beings rather than automatons? We start from the principle that all mental processes, in particular consciousness, correspond to electrophysiological processes in the nerve cells of the brain, that, in fact, consciousness is just the inner perception of some of these electrophysiological processes. If, according to the crude physical determinism referred to above, one could deduce our actions from the approximately known initial state of our nerve cells, the impossibility of free will could be proven in a physical-empirical sense, at least in principle. But since we now know that the slightest, immeasurably small differences in the initial state can lead to completely different final states (that is,

decisions), physics cannot empirically prove the impossibility of free will. However, we haven't gotten rid of Kant's antinomy on the impossibility of freedom.
We have refuted the crude determinism of working physics but not the exact determinism of Laplace. We lose the latter when we follow a causal chain backwards to ever more finely differentiated causes, losing it at that point where the theory of cognition asks ontological questions concerning the limits of mathematics itself to give an exact replica of reality. I believe that such a limit exists; to make the argument, I must digress a bit.
The immense successes of the mathematically exact sciences have led particularly the physicists to believe that empirically observable reality must obey the rules of mathematics down to the last detail. That physical reality does obey mathematical laws is one of the most amazing and wonderful discoveries that mankind has made. Mathematics, however, is not an experimentally deducible science, rather it is brought forth from the human intellect, as Athena was brought forth from the head of Zeus.
Mathematical knowledge is conclusive, i.e. its truthcontent is communicable, but it is knowledge *a priori*. When a physicist uses this knowledge to make predictions based on a few measurements and an appropriate theory, about natural phenomena at some time and place completely different from where he or she made the original measurements, and these predictions prove true, it borders on the miraculous. The physicist merely confirms with satisfaction that the theory must be right. But why should the world of objects subjugate itself at all to a theory, a mathematical structure?
Kant gave an ingenious answer to this question: it is our perception itself which structures reality in this way, that is, only that which is reflected as reality in our minds obeys mathematical rules; we know nothing about the world outside (or the "Ding an sich"). As clever as it is, I hold this idea to be incorrect

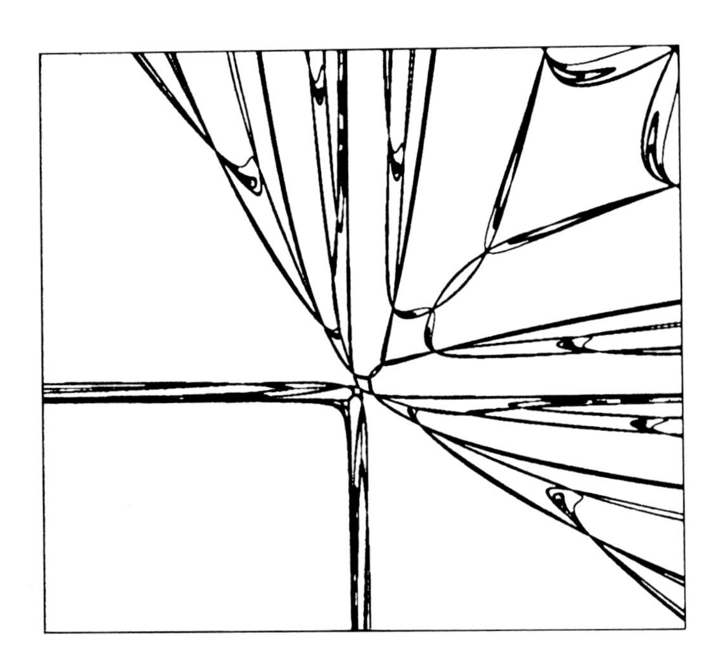

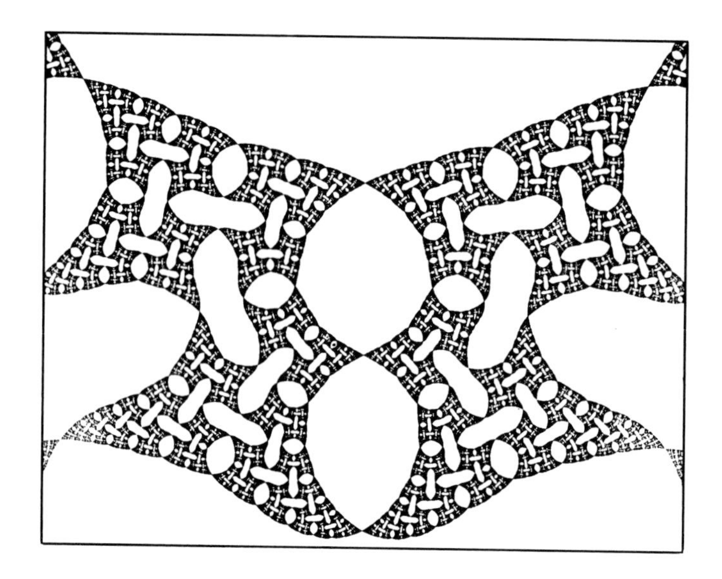

and accept the idea of an evolutionary theory of cognition, the foundations of which go back to the physicist Ludwig Boltzmann and which has been developed further especially since the work of Konrad Lorenz.

The basic idea is the following: it isn't our sensory and perceptual activity that forces nature into a strait-jacket of mathematics, it is Nature, which, in the process of our evolutionary development, has impressed mathematics into our reason as a real, existing structure, inherent to herself. Less abstractly: the ape, from which we are descended, had to have a very accurate idea of the geometry of space *actually existing* if he weren't to fall out of his tree and break his neck. Similarly, one can argue that the evolution of our abilities for abstraction and manipulation of logical symbols must be oriented on actually existent structures in the real world.

The abilities necessary to do mathematics are part of the genetically fixed experience of our species, *a priori* for the individual and *a posteriori* for the species as a whole.

Still, the enormous breadth of the mathematical description of nature is a miracle. Science has not yet reached a clearly discernable limit of this method, though I cannot rid myself of the suspicion that the peculiar paradoxes which arise in the *interpretation* of quantum mechanics may be indicators of such limits. This breadth is astounding because our mathematical ability (if the evolutionary theory of cognition is correct) was acquired by our ancestors through experiences with the relatively coarse structures and objects of the everyday world.

It is by no means obvious that our geometrical and logical powers should extend beyond the everyday world. That they do indeed do so indicates that reality itself is correspondingly structured (mathematically) to a much larger extent.

Even if these structural principles are extrapolated by ever deeper constructions and theorems, it is *simply improbable* that reality is completely and exhaustively mappable by mathematical constructs, from the greatest cosmological dimension down to the very last microscopic detail.

This is no easy admission for a theoretical physicist, but the evolutionary theory of cognition actually makes compulsory the assumption that the mathematical abilities of the species *homo sapiens* are in principle limited because of their biological basis, and, therefore, cannot completely contain all structures of reality. In other words, there must be limits to the mathematical description of nature.

Because of this, the determinism of Laplace cannot be absolute and the question of the possibility of chance and freedom is open again!

The pictures in this exhibition have another, completely different aspect – they simply are beautiful. The chaotic component shown in the very fine structures does not overpower the whole work; there are large areas of order, sustained by regularity, and chaos and order appear to be joined in harmonious balance.

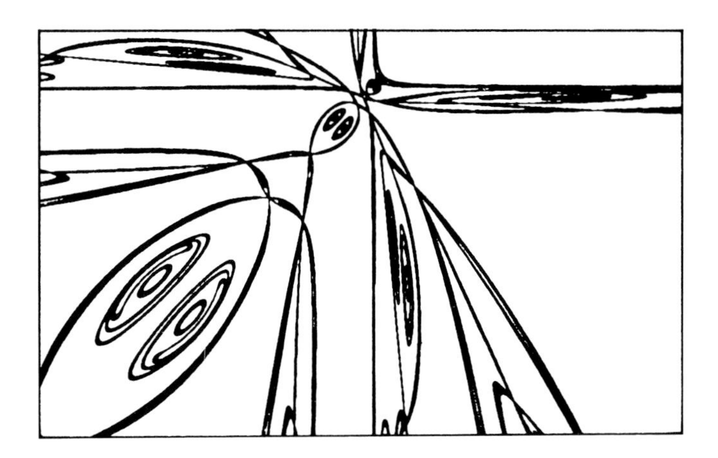

Precisely this *mixture* of order and disorder is fascinating, and, what is crucial to these new insights, *typical* for natural processes. Here, the science of dynamical systems provides an answer to a second, emotional question: why is it that the products of our technology, the entire technical world, seem to be *un*natural when they are products of *natural* science?

Why is it that the silhouette of a storm-bent, leafless tree against an evening sky in winter is perceived as beautiful, but the corresponding silhouette of any multi-purpose university building is not, in spite of all efforts of the architect?

The answer seems to me, even if somewhat speculative, to follow from the new insights into dynamic systems. Our feeling for beauty is inspired by the harmonious arrangement of order and disorder as it occurs in natural objects – in clouds, trees, mountain ranges or snow crystals. The shapes of all these are dynamical processes jelled into physical forms, and particular combinations of order and disorder are typical for them.

In comparison, our technical products are made stiff by the complete orderliness of their forms and functions, and all the more so the more perfect they are. Such *total* regularity does not *contradict* laws of nature, but we now know that it is *untypical* for even quite "simple" natural processes. We are dealing here with artificially produced borderline cases of nature, *pathological cases*, if you wish.

One can ask if this last observation is really so surprising – doesn't any unspoiled viewer see this right away? Very true, but we scientists (if I may include my colleagues) were not *unspoiled* viewers! We had built up our concepts (and prejudices) about the characteristic behavior of natural systems by observing artificial systems, systems which were *chosen* precisely for their regularity! This total regularity was a prerequisite for the mathematical description of the process. Only the advent of powerful computers freed us from this restriction. Computers – generally suspected of impressing total order and discipline on every facet of life – have made possible this better understanding of harmony and chaos.

That is part of the excitement surrounding these pictures: they demonstrate that out of research an inner connection, a bridge, can be made between rational scientific insight and emotional aesthetic appeal; these two modes of cognition of the human species are beginning to concur in their estimation of what constitutes nature. Even more: science and aesthetics are in agreement about what is actually missing in technical objects as compared to natural objects: it is the luxury of an appropriate portion of irregularity, disorder, and unpredictability. This insight could help us in a very basic way to give technology, on which we are more and more dependent for survival, a humane face.

# Refractions of Science into Art

HERBERT W. FRANKE

Art critics in the centuries to come will, I expect, look back on our age and come to conclusions quite different than our own experts. Most likely the painters and sculptors esteemed today will nearly have been forgotten, and instead the appearance of electronic media will be hailed as the most significant turn in the history of art. The debut of those first halting and immature attempts to achieve that ancient goal, namely the pictorial expression and representation of our world, but with a new media, will finally be given due recognition.
It will be pointed out that back then (now!) it became possible for the first time to create three dimensional pictures of imaginary landscapes and other scenes with photographic precision, and with these pictures not just to capture an instant in time but to include the reality of change and movement. Perhaps this is the most important aspect of the new turn: the time dimension has been unlocked for pictures, and planar or three dimensional scenes in perspective, even from points of view not accessible to the human eye or camera, can be arranged freely.
Today, however, artists using the computer find themselves forced to the periphery of the art "scene". In part formal criteria are used to support their rejection, in part the objections come from concrete commercial interests.

## *The Visual Age*

The computer is an instrument of data processing, and the concept of "data" seems to denote numbers, not pictures. In actual fact, however, pictures are just another means of describing content; they can always be coded as numbers and thus may be processed by a computer. The results are called computer generated graphics, computer graphics for short. Originally a sidetrack of computer applications, they have in the meantime grown immensely in significance.
Of course computer graphics were not initially developed with artistic goals in mind; they were the result of practical considerations in science, technology, and commerce. The realisation that pictures are easier to understand than long lists of numbers led to the replacement of numbers by block diagrams, curve plots, and so forth, in all these fields. As a consequence, computer graphics were developed. The first rather inadequate means of realizing this transition was the printer, which nevertheless showed the effectiveness of pictorial presentations so convincingly that mechanical drawing instruments, called plotters, were developed almost as a matter of course. The plotter uses the principle of a servocontrolled pen which draws on either a drum or a plane surface. This makes possible the mechanical production of line drawings precise enough for, say, architectural designs, cartography, and so forth.

One disadvantage to be noted however, is the extreme slowness of mechanically moved systems as compared to the electronic processing speed of the computer. Configurations which the computer can calculate in fractions of a second can take minutes, sometimes hours, to plot out.
The first real breakthrough in computer graphics came with the electronic plotter, the display tube, which shows visual information in both the same manner and at the same speed as the TV tube. Finally a display medium was available which was fast enough to keep up with the processing speed of the computer. It also made possible the generation of moving scenes and even interactive graphics.
One of the primary uses of computer graphics is CAD, computer automated design; to take just one example, the construction drawings and plans of component parts stored in the computer can be translated automatically for the control of appropriate machine tools.
New techniques of picture processing were developed from older methods of picture analysis and pattern recognition based on methods from photography. These techniques allow images acquired from science, technology, and medicine to be more easily evaluated, and in some cases, to be possible for the first time.
Some mathematicians and programmers used the aesthetic possibilities of graphics system from the very beginning in the early sixties. Most avoided using the word "art" in connection with their work, however, thus dodging a conflict with the art establishment. But some did stand up to the critics and acknowledged the computer as a new tool of the visual arts, calling their method "computer art". This caused many heated discussions about whether or not it is at all possible for art to be created with the help of a machine, and there were plenty of theoretical arguments opposing the idea. Happily, the representatives of this new direction were little influenced by these discussions and continued their work independently of any theoretical objections. Indeed, an impressive repertoire of computer graphics has accumulated, and this collection reflects the technical progression from simple drum plotter to high-resolution visual display graphics.

## *The Picture as a Means of Expression*

Some specialists are worried about the transition from verbal language to pictures, as if visual information was basically inferior to verbal or aural information. This position is untenable; on the contrary, the proverb, "a picture is worth a thousand words", is supported by arguments from the psychology of perception. Not only can the visual system handle an information flow approximately ten times that of all other sensory systems together, but the information processed can have a two, and to some extent even a three dimensional structure. Besides offering an additional system of encoding information, this change from words to pictures has shifted the basis of the descriptive possibilities we use to grasp our world.
If we employ language as a means of communication, a linear medium arranged as a time-series, we automatically favor linear ordering principles, e.g.,

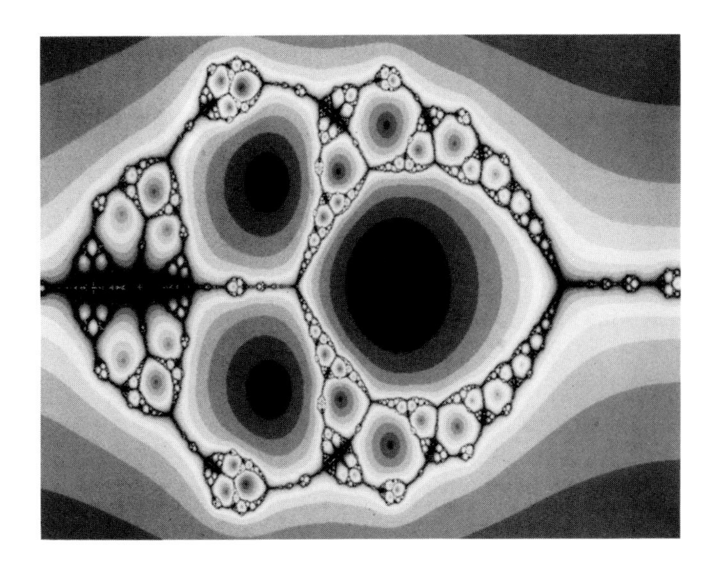

causality or historical process. Visual languages allow us on the other hand to see those very important connections which manifest themselves as loop processes, interactions, communication networks, and so forth. Perhaps our inability to think in terms of networks is due in no small measure to our restriction to the descriptive system of verbal language.
Such reflections show us that the statements of various opponents of visualisation are not necessarily sound. If they assert that pictures leave much less room for imagination than a written text, they can be refuted. Even comics need a certain amount of creative completion by their readers (shown by the fact that some parents cannot understand them). Pictures do indeed fix one kind of information more strongly than words do, but enough gaps are left for the viewer to fill. In comics, for example, the logic of the plot, the social relationships of characters depicted, any emotional background, is all supplied by the fantasy of the viewer. The same is true for other visual media, for film, television, video and others.
Unfortunately, in discussions about new computer pictures such banal applications as commercials gain too prominent a position. A basic problem, more important today than ever before, is missing in these discussions, namely, how to cast the complex interrelations of our modern world into an understandable form, how to make them conceivable and comprehensible. The more complex the situation is, the more appropriate a pictorial representation becomes.
One of the fundamental uses of computer graphics, particulary interactive graphics, is in science and education. Computer generated pictures are used to present the descriptive content of a great variety of mathematical formulas in a more striking and comprehensive way than has ever been known. They confirm the idea that coding information in a two and, to a certain degree, three dimensional form will show connections that cannot be imagined from the formulas alone. Beyond this, the computer offers us the possibility of experimentation; the mathematician can check the influence of parameters,

the result of transformations, the limiting values of iteratively applied calculations, the physicist can check the effect of boundary conditions on fields, and the chemist can analyse the spatial structure of aggregates of molecules.

## *Is it Art?*

Comprehensiveness and clear impression: how do these come about? The psychology and aesthetics of information theory point the way. For instance, a harmonious arrangement, balance, symmetry, all ordering principles in the sense of information theory, are means to this end.
That these concepts can apply to visual phenomena is understandable, for the same Gestalt principles are employed that the human brain uses to recognize, analyse, and interpret objects. At the same time, these are the same formal relationships associated with the ideal of classical beauty. If any collection of data is presented in such a way that it attains optimal visual recognition, e.g. with the help of a computer, then there will be a simultaneous convergence with the characteristics of the Beautiful.
Modern art studies have shown, however, that meeting the classical definition of beauty is not in itself sufficient to create a work of art. In addition there must be something to stimulate interest, demand involvement, and motivate further thoughts. The psychology of information theory proves that such stimuli emanate particulary from "innovation", i.e. from the confrontation of our visual apparatus with that which seems new, which has not occured before.
Impressibility and innovation – these are the interfaces between the pictures of science and those of art. A scientific photograph, for example, is not a direct expression of reality. Rather it is the result of a complicated optical or even electronic conversion, a conversion with the goal of the best possible visual recognition in mind. What is attained by the use of stains or polarization filters in microscopy has nothing to do with the real colors of the object. They serve only to make certain regions more distinguishable.
Of course the objects themselves, like crystalline structures, can fulfill the prerequisites of form necessary for feature recognition. But only through a conversion with the methods of microscopy can those pictures be produced that show such astonishing resemblance to works of art.
Nature's art forms – this concept of Ernst Haeckel's is an expression of the striking relationship between two seemingly disparate areas of visual experience. Wandering across the surface of a crystal with the help of a microscope can truly be an aesthetic pleasure. Again and again new regions come into view reminding one of landscapes, and there is always the possibility of increasing the magnification, thus making exploration into further layers of structure possible.
A surprising similarity to microscopy is found in the work exhibited here and the research team producing it. Their name "Complex Dynamics" indicates the scientific nature of their activity, but, significantly, they have coined the word "MAP ART" for their graphics. This synthetic name is an expression of the same experience made by biologists and crystallographers working with

microscopes, and subsequently other natural scientists, from particle physicists to astronomers.
One striking difference exists between natural scientists and mathematicians. While it is possible to say that the scientific photographer draws from structures in nature, that is, he passes off in some sense copies of nature as works of art, no such accusation can be leveled at the mathematician. The employment of the computer may superficially resemble the use of a microscope, but what a computer produces is not prescribed by nature, it is something created entirely by the human mind. Thus, it is not just a very successful visual presentation of nature which the team "Complex Dynamics" has achieved, indeed they have created the very content of their presentations. This fact must be emphasized, that the pictures communicate genuine innovation and indicate problems, such as complexity, which are fruitful to consider and pursue. One could of course object that the themes are unusual for art, that they are themes which only interest scientists and technologists, and so can only be considered art for a scientific-technical subculture. This objection ignores the fact that the problematics of complex systems are found at the very basis of biological life and social systems, in the origin of the universe as well as in philosophy. So indeed it is art according to all the usual definitions and could only be disqualified specifically on the grounds that a machine was used, the computer.

### *The Expansion of Computer Art*

Art has much to do with communication, communication between the artist and his audience. In the optimal case, this is a closed cycle; the artist presents his work to the audience, he provokes a reaction which he takes as feedback, and, perhaps with the intention of making himself more understandable, he incorporates this response into his further work.

What obstacles are there to the spreading of computer graphics and computer art? As long as they were only considered drawings from the plotter, the main problem was the uncertainty of the experts, the art historians and the critics, and above all the gallery owners. The problem was that the computer can produce an arbitrary number of equally good "originals", which can be a detriment in the business world of art. Since then it has been discovered that some of the pioneer work produced by computers was completely unique and irreproducible. It would have been worth collecting then!

This situation is much more difficult now, when the pictures no longer are drawn out on the plotter but are created on the display monitor. From here they have to be recorded, for example with an undistorted, computer controlled photograph. But this immediately raises old reservations and revives the discussion about whether a photograph can be a work of art.

There are other means of transferring pictures from display monitors on to paper in an incontestable way, in particular by the use of ink-jet printers. Unfortunately, such techniques are still only available to a few users, and so most modern results, even when three dimensional, perspective images are involved, are on slides or photopaper. The situation is especially difficult for moving pictures, which cannot be presented in galleries or by conventional art dealers. In order to show these works to the public, completely new avenues must be opened, for example those institutions which deal in television pictures, like video cassette libraries, or television stations. Is there simply too little demand for moving, computer generated sequences, or is it sluggishness on the part of the media, an unwillingness to switch to a new track?

What is the best way for a computer-artist to bring his works before the public? The easiest method seems to be to accept the modalities of galleries, and to bring out editions of static pictures transferred onto paper by the silk screen process. When these creations have been accepted, possibilities should open up for moving sequences of pictures.

## *Art and Technology Today*

Clever business strategies and an excellent product do not help if the public rejects the product. One reason for the rejection of computer art could be that it is being produced with the help of a highly technical medium, a medium very much under cross-fire today. Interceding in favor of technology as a legitimate means of creating art, means confronting the question of whether or not, in a world where technological progress itself has become dubious, art at least should be kept free of machines. The answer is simple. The technology necessary for computer art is essentially different than the technology used by the energy and manufacturing industries, those technologies mainly under criticism. The computer is a product of information technology and consumes only a minimum of energy and material resources. There is no reason to confine employment of data processing systems to science, technology, business, and administration, and exclude artistic spheres.

It is not a question of replacing the conventional methods of artistic creation with electronics or a machine. Rather it makes sense to use every means possible to extend the range of artistic expression. The art of every age has used the means of its time to give form to artistic innovation. This has never been purely for technical or practical reasons, it is also determined by the communicative behavior of the public. Why shouldn't the computer, that universal medium of information and communication which has even invaded our private homes, be used as a medium and instrument of art?

# DO IT YOURSELF

From the reactions to our previous publications we know that many students are less interested in our pictures as such than in trying to redo them on their own equipment. This is of course exactly the kind of response that we intend to elicit. To help achieve this goal we shall now give a few hints for all those who find it too cumbersome to digest the main text.
We show in two examples how you may produce typical pictures of this book. Remember: we are studying feedback processes where the $k$-th point $(x_k, y_k)$ generates the $(k+1)$-st point $(x_{k+1}, y_{k+1})$ by means of a given law:

$$x_{k+1} = f(x_k, y_k; p),$$
$$y_{k+1} = g(x_k, y_k; q), \qquad (1)$$

$p$ and $q$ are parameters that we keep constant during each iteration $(x_0, y_0) \mapsto (x_1, y_1) \mapsto \ldots \mapsto (x_k, y_k) \mapsto \ldots$
There are two different types of representation. One is to consider the $(x,y)$-plane at fixed values of $p$ and $q$, and to analyse each point $(x,y)$ for its dynamical relation to the respective attractor. This way we explore the structure of the domains of attraction, and of their boundaries, the *Julia sets*.
The other possibility is to choose a fixed point $(x,y)$ and to follow its fate at different values of the parameters. The results are recorded, point by point, in the $(p,q)$-plane. This produces pictures of the *Mandelbrot set* kind.

*Experiment 1. Basins of Attraction and Julia Sets*

Consider, as an example, the complex feedback process $z \mapsto z^2 + c$. Decompose the complex numbers $z$ and $c$ into their real and imaginary parts, $z = x + iy$, $c = p + iq$. The process is then described by a law of the form (1):

$$x_{k+1} = x_k^2 - y_k^2 + p,$$
$$y_{k+1} = 2x_k y_k + q. \qquad (2)$$

Since infinity is always an attractor for this process we set ourselves the goal to color its domain of attraction. The colors are to indicate how long it takes a point $(x,y)$ to escape towards infinity.
Assume the monitor has a graphical resolution of $a$ times $b$ points. Let there be $K+1$ colors that can be displayed simultaneously, numbered 0 through $K$. Color 0 is black.

*Step 0:*
Choose a parameter $c = p + iq$. Choose $x_{min} = y_{min} = -1.5$, $x_{max} = y_{max} = 1.5$, $M = 100$.
Set $\Delta x = (x_{max} - x_{min})/(a-1)$, $\Delta y = (y_{max} - y_{min})/(b-1)$.
For all pixels $(n_x, n_y)$ with $n_x = 0, \ldots, a-1$ and $n_y = 0, \ldots, b-1$, we go through the following routine.

*Step 1:*
Set $x_0 = x_{min} + n_x \cdot \Delta x$, $y_0 = y_{min} + n_y \cdot \Delta y$, $k = 0$.

*Step 2* (iteration step):
Calculate $(x_{k+1}, y_{k+1})$ from $(x_k, y_k)$ using the law (2). Increase the counter $k$ by 1 ($k := k+1$).

*Step 3* (evaluation step):
Calculate $r = x_k^2 + y_k^2$.
(i) If $r > M$ then choose color $k$ and go to step 4.
(ii) If $k = K$ then choose color 0 (black) and go to step 4.
(iii) $r \leqq M$ and $k < K$. Go back to step 2.

*Step 4:*
Assign color $k$ to the pixel $(n_x, n_y)$ and go to the next pixel, starting with step 1.

*Remarks*

It pays to allow for large $K$, say $K = 200$. If the number of different colors is limited, then it is helpful to use them in a periodic manner. Even two colors (black and white) can produce interesting pictures this way.
The computing time can be reduced by a factor of 2 by using the symmetry of the process (2). Both points $(x,y)$ and $(-x, -y)$ give the same result, thus the picture is symmetric with respect to the origin.
Varying the entries $x_{min}$, $x_{max}$, ... in step 0 you may produce blow-ups.
In step 4, everything will come out black that does not converge towards infinity after $K$ steps. This includes the domains of other attractors if they exist. In case of process (2) a second attractor exists whenever $c = p + iq$ is taken from inside the Mandelbrot set (see experiment 2, below). If you want this second domain to be colored in a corresponding manner, you must find the attractor first. For that purpose, iterate the "critical point" $(x_0, y_0) = (0,0)$ for a while and observe where it goes. But watch out: the attractor need not be a single point, it may consist of a cycle. Then choose a point from this cycle, define a small disk around it, and ask in step 3(i) whether or not the iteration has hit the disk.
If you want to produce pictures like Maps 64–66, choose a polynomial $p(z)$ and let its zeroes become attractors by application of Newton's root finding algorithm to the problem $p(z) = 0$:

$$z_{k+1} = z_k - r \frac{p(z_k)}{p'(z_k) + s \cdot i}. \tag{3}$$

(The standard algorithm has $r = 1$ and $s = 0$, but other choices in the ranges $0 < r < 2$, $0 < s < 1$ give interesting results. You may also choose complex $r$.)

*Problem:*
Choose $p(z) = z^3 - 1$ so that $z = 1$, $z = -1/2 \pm i\sqrt{3}/2$ are the three attractors. Decompose (3) into real and imaginary parts (for $s = 0$, e.g.) and color the three domains of attraction.

*Experiment 2. The Mandelbrot Set*

According to Fatou, the process (2) cannot have a second attractor besides infinity whenever the "critical point" $z_0 = 0$ converges towards infinity. (In general, $z_0$ is a critical point for the process $z \mapsto f(z)$ if $f'(z_0) = 0$. This is the basis of the following recipe in which the Mandelbrot set is coming out black with colored surroundings.
Let the monitor have $a$ times $b$ points again and $K+1$ colors. Color 0 be black.

*Step 0:*
Choose $p_{\min} = -2.25$, $p_{\max} = 0.75$, $q_{\min} = -1.5$, $q_{\max} = 1.5$, $M = 100$.
Set $\Delta p = (p_{\max} - p_{\min})/(a-1)$, $\Delta q = (q_{\max} - q_{\min})/(b-1)$.
For all points $(n_p, n_q)$ of the monitor $(n_p = 0, \ldots, a-1)$, $(n_q = 0, \ldots, b-1)$ go through the following routine:

*Step 1:*
Set $p_0 = p_{\min} + n_p \cdot \Delta p$, $q_0 = q_{\min} + n_q \cdot \Delta q$, $k = 0$, $x_0 = y_0 = 0$.

*Step 2:*
Calculate $(x_{k+1}, y_{k+1})$ by means of (2). Then set $k := k+1$.

*Step 3:*
Calculate $r = x_k^2 + y_k^2$.
(i) If $r > M$ then choose color $k$ and go to step 4.
(ii) If $k = K$ then choose color 0 (black) and go to step 4.
(iii) $r \leqq M$ and $k < K$. Repeat step 2.

*Step 4:*
Assign color $k$ to point $(n_p, n_q)$ and go to the next point (step 1).

Remarks 1–3 of experiment 1 apply respectively. You will notice that experiment 1 produces most satisfying results when $c = p + iq$ is chosen from the neighborhood of the boundary of the Mandelbrot set. Try blow-ups at this boundary, in the $(p,q)$-plane. But note: the closer you are to the boundary the more iterations it takes to decide whether or not the critical point escapes to infinity.

# DOCUMENTATION

As a guide for those who want to understand our pictures in detail, we give here the necessary information. For all pictures that have not been identified in the text we specify

- the underlying process,
- the parameter choice,
- the window of the displayed variables.

Note that the window can only approximately be given because in many cases the original had to be clipped or turned around for lay-out reasons.

## *The Process* $x \mapsto x^2 + c$

### x-Plane Pictures for Different c

| Fig. | Map | $c$ | | Re $x$ | Im $x$ |
|---|---|---|---|---|---|
| 3 | | −0.12375 | +0.56508$i$ | −1.8 ... 1.8 | −1.8 ... 1.8 |
| 4 | | −0.12 | +0.74$i$ | −1.4 ... 1.4 | −1.4 ... 1.4 |
| 6 | | −0.481762 | −0.531657$i$ | −1.5 ... 1.5 | −1.5 ... 1.5 |
| 7 | | −0.39054 | −0.58679$i$ | −1.5 ... 1.5 | −1.5 ... 1.5 |
| 8 | | 0.27334 | +0.00742$i$ | −1.3 ... 1.3 | −1.3 ... 1.3 |
| 9 | | −1.25 | | −1.8 ... 1.8 | −1.8 ... 1.8 |
| 10 | | −0.11 | +0.6557$i$ | −1.5 ... 1.5 | −1.5 ... 1.5 |
| 11 | | 0.11031 | −0.67037$i$ | −1.5 ... 1.5 | −1.5 ... 1.5 |
| 12 | | $i$ | | −1.5 ... 1.5 | −1.5 ... 1.5 |
| 13 | | −0.194 | +0.6557$i$ | −1.5 ... 1.5 | −1.5 ... 1.5 |
| 14 | | −0.15652 | +1.03225$i$ | −1.7 ... 1.7 | −1.7 ... 1.7 |
| 15 | | −0.74543 | +0.11301$i$ | −1.8 ... 1.8 | −1.8 ... 1.8 |
| | 18 | 0.32 | +0.043$i$ | −2 ... 2 | −1.5 ... 1.5 |
| | 20 | −0.12375 | +0.56508$i$ | −2 ... 2 | −1.5 ... 1.5 |
| | 22,25 | −0.39054 | −0.58679$i$ | −1.5 ... 1.5 | −1.5 ... 1.5 |
| | 24 | −0.11 | +0.67$i$ | −2 ... 2 | −1.5 ... 1.5 |

### c-Plane Pictures: Mandelbrot Set and Close-Ups

| Map | Re $c$ | | Im $c$ | |
|---|---|---|---|---|
| 26 | −2.25 | ... 0.75 | −1.5 | ... 1.5 |
| 27,52–54 | −0.19920 | ... −0.12954 | 1.01480 | ... 1.06707 |
| 29 | −0.95 | ... −0.88333 | 0.23333 | ... 0.3 |
| 30 | −0.713 | ... −0.4082 | 0.49216 | ... 0.71429 |
| 33,51 | −1.781 | ... −1.764 | 0. | ... 0.013 |
| 36 | −0.75104 | ... −0.7408 | 0.10511 | ... 0.11536 |
| 38 | −0.74758 | ... −0.74624 | 0.10671 | ... 0.10779 |
| 40 | −0.746541 | ... −0.746378 | 0.107574 | ... 0.107678 |
| 42,100,101 | −0.74591 | ... −0.74448 | 0.11196 | ... 0.11339 |
| 44,46,99 | −0.745538 | ... −0.745054 | 0.112881 | ... 0.113236 |
| 45,47 | −0.745468 | ... −0.745385 | 0.112979 | ... 0.113039 |
| 48,50 | −0.7454356 | ... −0.7454215 | 0.1130037 | ... 0.1130139 |
| 49 | −0.7454301 | ... −0.7454289 | 0.1130076 | ... 0.1130085 |
| 58–60 | −1.254024 | ... −1.252861 | 0.046252 | ... 0.047125 |

◁ *This picture was produced in collaboration with the IBM labs at Böblingen, FRG (W. Hehl, D. Wollschläger). It shows the effect of increased resolution. The calculation was done on an IBM 4361-5 with extended precision; the printer IBM 4250 gave a resolution of 4096 × 5120 points*

*Models for Magnetism*

Two hierarchical lattices with Potts spins are analyzed. The renormalization transformation in terms of the "temperature" variable $x = \exp(J/k_B T)$ reads

$$x \mapsto \left(\frac{x^2+q-1}{2x+q-2}\right)^2 \qquad \text{(Model I)}$$

$$x \mapsto \left(\frac{x^3+3(q-1)x+(q-1)(q-2)}{3x^2+3(q-2)x+q^2-3q+3}\right)^2 \qquad \text{(Model II)}$$

where $q$ is the number of Potts states in the original model but is treated as a complex parameter here.

## x-Plane Pictures for Different q

| Fig. | Map | Model | $q$ | Re $x$ | Im $x$ |
|---|---|---|---|---|---|
| 53a | | I | $-1.0$ | $-6$ ... 4 | $-5$ ... 5 |
| b | | I | $-0.1$ | $-6$ ... 4 | $-5$ ... 5 |
| c | | I | 0 | $-6$ ... 4 | $-5$ ... 5 |
| d | | I | 1.0 | $-5$ ... 5 | $-5$ ... 5 |
| e | | I | 1.2 | $-5$ ... 5 | $-5$ ... 5 |
| f | | I | 1.6 | $-5$ ... 5 | $-5$ ... 5 |
| g | | I | 2.0 | $-5$ ... 5 | $-5$ ... 5 |
| h | | I | 2.5 | $-5$ ... 5 | $-5$ ... 5 |
| i | | I | 2.9 | $-5$ ... 5 | $-5$ ... 5 |
| j | | I | 3.0 | $-5$ ... 5 | $-5$ ... 5 |
| k | | I | 3.1 | $-5$ ... 5 | $-5$ ... 5 |
| l | | I | 4.0 | $-4$ ... 6 | $-5$ ... 5 |
| | 3 | I | 4 | $-5.5$ ... 7.9 | $-5$ ... 5 |
| | 4 | I | $-0.1$ | 0.42 ... 1.98 | 2.5 ... 3.67 |
| | 5 | I | $1.09582+2.07142i$ | $-0.5$ ... 1.5 | $-1.45$ ... 0.7 |
| | 6 | I | $1.21 \quad +0.01i$ | $-2.1$ ... $-0.3$ | $-0.5$ ... 0.5625 |
| | 7 | II | 2 | $-15$ ... 15 | $-11$ ... 11 |
| | 8–10 | II | $1.2 \quad +2i$ | $-3.2$ ... 3.8 | $-3.3$ ... 1.9 |

## q-Plane Pictures

In both models, the points $x=1$ and $x=\infty$ are always superstable attractors of the renormalization process, corresponding resp. to infinite real temperature (complete disorder) and zero real temperature at $J>0$ (complete magnetic order). In addition, there may be one or two extra attractors because there are two free critical values $x_1=0$ and $x_2=(1-q)^2$. Our parameter studies are records of the time it takes these critical values to get close to $x=1$ (one main color) or $x=\infty$ (other main color). With $q$ inside the black areas, the critical points converge towards one of the extra attractors.

Note that in case of model I, the two critical values 0 and $(1-q)^2$ have a complementary fate so that the parameter charts they produce look identical

| Fig. | Map | Model | crit. value | Re $q$ | Im $q$ |
|---|---|---|---|---|---|
| 54 b | | I | 0 or $(1-q)^2$ | 1.85 ... 2.15 | 1.5 ... 1.8 |
| c | | I | 0 or $(1-q)^2$ | −1. ... 3.5 | −2.25 ... 2.25 |
| d | | II | $(1-q)^2$ | −1. ... 2.65 | −2. ... 2. |
| e | | II | 0 | 1.92 ... 1.97 | 0.88 ... 0.93 |
| f | | II | 0 | −0.8 ... 3.5 | −2. ... 2. |
| | 1 | I | 0 or $(1-q)^2$ | −1.5 ... 4.5 | −2.25 ... 2.25 |
| | 2 | I | 0 or $(1-q)^2$ | 1.86 ... 2.21 | 1.51 ... 1.77 |
| | 11, 12 | I | 0 or $(1-q)^2$ | 1.2882 ... 1.2963 | 0.9695 ... 0.9753 |
| | 13 | I | 0 or $(1-q)^2$ | 1.290681 ... 1.291136 | 0.97277 ... 0.973098 |
| | 14, 15 | II | 0 | 1.9116 ... 1.9784 | 0.88 ... 0.93 |
| | 16 | II | 0 | −1.2 ... 3.2 | −1.65 ... 1.65 |

## *Newton's Algorithm*

Given the polynomial $p(z)$, the following modified Newton algorithm is used to find the roots of $p(z)=0$:

$$N(z)=z-\frac{p(z)}{p'(z)+s\cdot i}.$$

| Map | $p(z)$ | $s$ | Re $z$ | Im $z$ |
|---|---|---|---|---|
| 61–63 | $(z-1)(z+\frac{1}{2})(z^2+1)$ | −0.5 | Representation on Riemann sphere | |
| 64, 65 | $(z-1)(z+\frac{1}{2})(z^2+1)$ | −0.5 | −0.35 ... 0.52 | −0.91 ... 0.22 |
| 66 | $z^3-1$ | 0 | −1.333 ... 1.333 | −1 ... 1 |
| 75, 76 | $(z-1)(z^2+z+\frac{5}{4})$ | 0.25 | Representation on Riemann sphere | |
| 77, 78 | $(z-1)(z^2+z+\frac{1}{2})$ | 0 | −2.46 ... 1.71 | −0.28 ... 0.28 |
| 89–98 | For $0\leq\alpha\leq1$ we have investigated the rational functions: $z\mapsto z+\alpha\left(\frac{p_1(z)}{p_1'(z)+0.25i}\right)+(1-\alpha)\left(\frac{p_2(z)}{p_2'(z)-0.5i}\right)$, where $p_1(z)=(z-1)(z^2+z+1.25)$ and $p_2(z)=(z-1)(z+0.5)(z^2+1)$. | | | |

## *Processes in Two Real Variables*

Lotka-Volterra Equations Discretized by Means of Modified Heun Algorithm (see Section 8)

Maps 79–82: $h=k=0.739$,
Maps 83–86: $h=0.8$, $k=0.86$.
All maps display the window $0\leqq x_1\leqq 6$, $0\leqq x_2\leqq 4.5$.

Boundary Value Problems Discretized on N Points [see Section 7 (7.2, 7.10)]

*a)* $f(u)=u-u^2$, $N=2$

| Map | $\mu$ | $h$ | $x_1$ | $x_2$ |
|---|---|---|---|---|
| 17 | 2.1 | 1.7 | $-5$ ... 5 | $-5$ ... 5 |
| 67 | 2.1 | 0.1 | $-5$ ... 5 | $-5$ ... 5 |
| 19, 68 | 3.2 | 0.2 | $-5$ ... 5 | $-5$ ... 5 |
| 23, 71–73 | 3.2 | 1.9 | $-5$ ... 5 | $-5$ ... 5 |
| 21, 87 | 3.2 | 1.8 | $-6.68$ ... 6.68 | $-5$ ... 5 |
| 69, 70, 88 | see Section 7 (7.14) | | | |

*b)* The last pictures to be explained Maps 55–57, 74 belong to the series of discretized boundary value problems, with the following special choices: $f(u)=u-u^3$, $N=6$, $\lambda=50$, $h=0.4$. With $N=6$, the problem is really 6-dimensional, but the five real solutions

$$(0, 0, 0, 0, 0),$$
$$\pm(0.63960, 0.89354, 0.96368, 0.96368, 0.89354, 0.63960),$$
$$\pm(0.47340, 0.57200, 0.27789, -0.27789, -0.57200, -0.47340),$$

span a 2-dimensional plane. This is taken as the $(x,y)$-plane with the $x$- and $y$-axis defined in such a way that the solutions become the points $(0,0)$, $\pm(1,0)$, and $\pm(0,1)$ respectively. Maps 55–57, 74 then show the window $-4\leqq x\leqq 4$, $-3\leqq y\leqq 3$.

# INDEX

H.-O. Peitgen, D. Saupe

# The Science of Fractal Images

By M. F. Barnsley, R. L. Devaney, B.B. Mandelbrot, H.-O. Peitgen, D. Saupe, R. F. Voss

With contributions by Y. Fisher, M. McGuire

1988. 142 figures in 277 parts, 39 color plates. XIII, 320 pages.
ISBN 3-540-96608-0

**Contents:** People and events behind the "Science of Fractal Images". - Fractals in nature: From characterization to simulation. - Algorithms for random fractals. - Color plates and captions. - Fractal patterns arising in chaotic dynamical systems. - Fantastic deterministic fractals. - Fractal modelling of real world images. - Fractal landscapes without creases and with rivers. - An eye for fractals. A unified approach to fractal curves and plants. - Exploring the Mandelbrot set. - Bibliography. - Index.

**The Science of Fractal Images** is a new book based on a SIGGRAPH'87 course given by these five recognized leaders in the field:

- Michael F. Barnsley: *Fractal Modelling of Real World Images*
- Robert L. Devaney: *Fractal Patterns Arising in Chaotic Dynamics*
- Heinz-Otto Peitgen: *Fantastic Deterministic Fractals*
- Dietmar Saupe: *Algorithms for Random Fractals*
- Richard F. Voss: *Fractals in Nature: Characterization, Measurement and Simulation*

In addition, the book contains a foreword and an important essay by Benoit Mandelbrot entitled: *Fractal landscape without creases and with rivers.*

This is a "how-to" book. It bridges the gap between the mathematical foundations of fractal geometry and the computer generation of fractal images. This exciting new field is making important contributions in areas such as advertising, animation, publishing, communications, and image processing. Anyone wishing to see more than beautiful pictures will find state-of-the-art algorithms and detail previously known only to specialists.

Springer-Verlag
Berlin Heidelberg New York
London Paris Tokyo Hong Kong

H. W. Franke

# Computer Graphics - Computer Art

Translated from the German by G. Metzger, A. Schrack

2nd revised and enlarged edition. 1985. 133 figures, some in color. XII, 177 pages.
ISBN 3-540-15149-4
(Originally published by Phaidon Press 1971)

**Contents:** Installations and Methods. - History of Computer Art. - Theoretical Foundations of Computer Art. - The Future of Computer Art. - Bibliography. - Name Index. - Subject Index.

Ever since the emergence of computers people have used these machines for creative, artistic work: to create graphics, animated pictures, to simulate visual scenes, to synthesize music and even poems. Although all kinds of arguments have been used to deny a connection between computers and art, it is obvious that the results of this controversially discussed computerized art have improved substantially, both in technical and in aesthetical terms. **Computer Graphics - Computer Art** describes concepts and tools for computer graphics, computer animation, picture processing, computer-aided design, computer music, computer choreography and computer poetry. Moreover, it gives a historical review from the origins of computer art to current developments.

Springer-Verlag
Berlin Heidelberg New York
London Paris Tokyo Hong Kong